Praise for *The ALL NEW D...*

"When the original *Don't Think of an E... ...* came out in 2004, Lakoff showed us that progressive Democrats voted on values, not issues—just like the right. Now, Dr. Lakoff is back to prevent a relapse of bad framing. *The ALL NEW Don't Think of an Elephant!* is a must read, every bit as important as the first edition. This time we have to train ourselves to think for the long term. Buy this book, memorize it, and teach it to your children. Progressives may be smart, but we don't communicate our ideas well. This book is the blueprint for how to do better."

—**HOWARD DEAN**, former chair of the Democratic National Committee and founder of Democracy for America

"Lakoff single-handedly convinced liberals of the importance of language in winning political battles. Now he's back to finish what he started."

—**MARKOS MOULITSAS**, founder and publisher, Daily Kos

"*The All New Don't Think of an Elephant!* is an indispensable tool for progressives—packed with new thinking on framing issues that are hotly debated right now, and new insights on how to reclaim the political debate on meaningful terms that can yield true progress, not just political gains."

—**JENNIFER M. GRANHOLM**, former governor of Michigan

Praise for the first edition, *Don't Think of an Elephant!*

"It's not enough that we have reason on our side. We also have to use words and images powerful enough to persuade others. Lakoff offers crucial lessons in how to counter right-wing demagoguery. Essential reading in this neo-Orwellian age of Bush-speak."

—**ROBERT REICH**, author, *Aftershock*, former US Secretary of Labor

"Fascinating insights into why progressives lose good causes and how they can start winning again. You will never listen to a political speech the same way after reading this book."

—**TINA BROWN**, cofounder, *The Daily Beast*

"Don't let anyone tell you that the words don't make a difference; they can evoke the best and the worst in us. Read this book and be part of transforming our political dialogue to support our highest ideals and speak to the hearts of Americans."

—**JOAN BLADES** and **WES BOYD**, MoveOn.org

"Ever wonder how the radical right has been able to convince average Americans to repeatedly vote against their own interests? It's the framing, stupid! *Don't Think of an Elephant!* is a pithy and powerful primer on the language of American politics, and a vital reminder that he who creates the political framework controls the picture that's put inside. It's also a detailed roadmap out of the mess we are in. Lakoff shows how progressives can reclaim the political narrative—and, in the process, change our country and our world for the better."

—**ARIANNA HUFFINGTON**, cofounder and editor-in-chief, *The Huffington Post*

"This is a pocket manifesto for those who still wonder how a small group of rich, powerful oligarchs tied together the shoelaces of the progressive movement. Read it once, and know why we are losing. Read it twice, and we can restore sanity to the world."

—**PAUL HAWKEN**, author of *Natural Capitalism*

"*Don't Think of an Elephant!* is a work of genius. As George Lakoff explains how the right has framed the notion of the political center, he presents both the most original and the most practical analysis of United States politics in many years."

—**GEORGE AKERLOF**, Nobel Prize winner in Economics

"George Lakoff's *Don't Think of an Elephant!* is a wonderful example of what happens when you combine a linguist's ear for the subtleties of language with an understanding of the complexities of modern politics and a commitment to progressive ideals. Whether you think of yourself as a liberal, a progressive, or simply someone with an interest in how political language works, this is a must read."

—**GEOFFREY NUNBERG**, University of California, Berkeley

"Progressives have a lot to learn about persuading swing voters to our cause, and there's no better teacher than George Lakoff. This readable text couldn't be more timely; it should be read widely and put to work before November!"

—**DANIEL ELLSBERG**, author of *Secrets: A Memoir of Vietnam and the Pentagon Papers*

"If you want to take back our country, you have to take back your community. If you want to take back your community, you need to take back the debate. This book, and the video that goes with it, are your essential tools. What the Bush administration has done for obfuscation, George Lakoff's work does for clarification."

—**CARL POPE**, former president, Sierra Club

★ ★ THE ALL NEW ★ ★
DON'T THINK OF AN ELEPHANT!

Previous Books By George Lakoff

★ Political Action ★

The Little Blue Book
Don't Think of an Elephant!
Thinking Points

★ Deep Politics ★

Moral Politics
Whose Freedom?
The Political Mind

★ Linguistics and Cognitive Science ★

Metaphors We Live By
Women, Fire, and Dangerous Things
Philosophy in the Flesh
Where Mathematics Comes From
More Than Cool Reason

★ ★ THE ALL NEW ★ ★
DON'T THINK OF AN ELEPHANT!
KNOW YOUR VALUES
AND FRAME THE DEBATE

GEORGE LAKOFF

Chelsea Green Publishing
White River Junction, Vermont

The first edition of this book was published as *Don't Think of an Elephant!*

Editor: Joni Praded
Project Manager: Patricia Stone
Copy Editor: Deborah Heimann
Proofreader: Eric Raetz
Designer: Melissa Jacobson

Printed in the United States of America.
First printing September, 2014.
10 9 8 7 6 5 4 3   16 17 18

Our Commitment to Green Publishing
Chelsea Green sees publishing as a tool for cultural change and ecological stewardship. We strive to align our book manufacturing practices with our editorial mission and to reduce the impact of our business enterprise on the environment. We print our books and catalogs on chlorine-free recycled paper, using vegetable-based inks whenever possible. This book may cost slightly more because it was printed on paper that contains recycled fiber, and we hope you'll agree that it's worth it. Chelsea Green is a member of the Green Press Initiative (www.greenpressinitiative.org), a nonprofit coalition of publishers, manufacturers, and authors working to protect the world's endangered forests and conserve natural resources. *The ALL NEW Don't Think of an Elephant!* was printed on paper supplied by Thomson-Shore that contains 100% postconsumer recycled fiber.

ISBN 978-1-60358-594-1 (paperback)—ISBN 978-1-60358-595-8 (ebook)

The Library of Congress Cataloging-in-Publication Data is available upon request.

Chelsea Green Publishing
85 North Main Street, Suite 120
White River Junction, VT 05001
(802) 295-6300
www.chelseagreen.com

MIX
Paper from
responsible sources
FSC® C013483

To Kathleen,
Whose insights illuminate every page

★ CONTENTS ★

★ PREFACE ★

Ten years ago, in 2004, when the first edition of this book appeared, hardly anyone had heard of or thought about, much less discussed, how social and political issues were framed. Framing was an unknown and undiscussed concept, outside of the academic field of frame semantics.

Don't Think of an Elephant! became a best seller and changed all that. Discussions of how issues are framed are now commonplace in the national media. Millions of people hear the word "frame" in a discussion of issues and understand, at least basically, what it means. That is a lot for one small book to have accomplished.

But *Don't Think of an Elephant!* had higher goals. At the time, the Republicans were doing a much better job at framing issues their way than the Democrats were. Republican framing superiority had played a major part in their takeover of Congress in 1994. I and others had hoped that, starting in 2004, a widespread understanding of how framing worked would allow Democrats to reverse the trend.

In the 2008 election, Barack Obama led a Democratic sweep of the White House and Congress, using far superior framing, as well as superior on-the-ground tactics—besides being a far superior candidate. I had hoped that the superior framing would continue.

It didn't. Almost immediately after Obama's inauguration in 2009, the Republicans regained framing superiority in public discourse, and that played a major role in the ascendancy of the Tea Party in Congress and in state houses throughout the nation. Now Republicans are setting their framing sights on the cities as well as the states.

What happened?

This tenth anniversary edition of *Don't Think of An Elephant!* will do more than just recap what framing is and how it works. The goal of this edition is to explain what happened, why the Democrats have gone back to losing framing wars, and what can be done about it.

That's a tall order. Let's get started. We will recap Framing 101 and then go on to Framing 102 and beyond.

George Lakoff
Berkeley, California
June, 2014

Reframing Is Social Change

We think with our brains. We have no choice. It may seem that certain politicians think with other parts of their anatomy. But they too think with their brains.

Why does this matter for politics? Because all thought is physical. Thought is carried out by neural circuits in the brain. We can only understand what our brains allow us to understand.

The deepest of those neural structures are relatively fixed. They don't change readily or easily. And we are mostly unconscious of their activity and impact.

In fact, about 98 percent of what our brains are doing is below the level of consciousness. As a result, we may not know all, or even most, of what in our brains determines our deepest moral, social, and political beliefs. And yet we act on the basis of those largely unconscious beliefs.

My field—cognitive science—has found ways to study unconscious, as well as conscious, modes of thought. As a cognitive scientist, my job is to help make the unconscious conscious, to find out and let the world know what is determining our social and political behavior. I believe that such knowledge can lead to positive social and political change. Why? Because what goes on in people's brains matters.

Do we have to go to the neural level to understand our politics? In some cases, yes. Diving that deep will be important, and we will discuss the brain when necessary. But, on the whole, the most important brain structures for our politics can be studied from the perspective of the mind. They are called "frames."

Frames

Frames are mental structures that shape the way we see the world. As a result, they shape the goals we seek, the plans we make, the way we act, and what counts as a good or bad outcome of our actions. In politics our frames shape our social policies and the institutions we form to carry out policies. To change our frames is to change all of this. Reframing is social change.

You can't see or hear frames. They are part of what we cognitive scientists call the "cognitive unconscious"—structures in our brains that we cannot consciously access, but know by their consequences. What we call "common sense" is made up of unconscious, automatic, effortless inferences that follow from our unconscious frames.

We also know frames through language. All words are defined relative to conceptual frames. When you hear a word, its frame is activated in your brain.

Yes, in your brain. As the title of this book shows, even when you negate a frame, you activate the frame. If I tell you, "Don't think of an elephant!," you'll think of an elephant.

Though I found this out first in the study of cognitive linguistics, it has begun to be confirmed by neuroscience. When a macaque monkey grasps an object, a certain group of neurons in the monkey's ventral premotor cortex (which choreographs actions, but does not directly move the body) are activated. When the monkey is trained *not* to grasp the object, most of those neurons are inhibited (they turn off), but a portion of the same neurons used in grasping still turn on. That is, to actively *not* grasp requires thinking of what grasping would be.

Not only does negating a frame activate that frame, but the more it is activated, the stronger it gets. The moral for political discourse is clear: When you argue against someone on the other side using their language and their frames, you are activating their frames, strengthening their frames in those who hear you, and undermining your own views. For progressives, this means avoiding the use of conservative language and the frames that the language activates. It means that you should say what you believe using your language, not theirs.

Reframing

When we successfully reframe public discourse, we change the way the public sees the world. We change what counts as common sense.

Because language activates frames, new language is required for new frames. Thinking differently requires speaking differently.

Reframing is not easy or simple. It is not a matter of finding some magic words. Frames are ideas, not slogans. Reframing is more a matter of accessing what we and like-minded others already believe unconsciously, making it conscious, and repeating it till it enters normal public discourse. It doesn't happen overnight. It is an ongoing process. It requires repetition and focus and dedication.

To achieve social change, reframing requires a change in public discourse, and that requires a communication system. Conservatives in America have developed a very extensive and sophisticated communication system that progressives have not yet developed. Fox News is only the tip of the iceberg. Progressives need to understand what an effective communication system is and develop one. Reframing without a system of communication accomplishes nothing.

Reframing, as we discuss it in this book, is about honesty and integrity. It is the opposite of spin and manipulation. It is about bringing to consciousness the deepest of our beliefs and our modes of understanding. It is about learning to express what we really believe in a way that will allow those who share our beliefs to understand what they most deeply believe and to act on those beliefs.

Framing is also about understanding those we disagree with most. Tens of millions of Americans vote conservative. For the most part they are not bad people or stupid people. They are people who understand the world differently and have a different view of what is right.

All Politics Is Moral

When a political leader puts forth a policy or suggests how we should act, the implicit assumption is that the policy or action is right, not wrong. No political leader says, "Here's what you should do. Do it because it is wrong—pure evil, but do it." No political leader puts forth policies on the grounds that the policies don't matter. Political prescriptions are assumed to be right. The problem is that different political leaders have different ideas about what is right.

All politics is moral, but not everybody operates from the same view of morality. Moreover, much of moral belief is unconscious. We are often not even aware of our own most deeply held moral views.

As we shall see, the political divide in America is a moral divide. We need to understand that moral divide and understand what the progressive and conservative moral systems are.

Most importantly, a great many people operate on different—and inconsistent—moral systems in different areas of their lives. The technical term is "biconceptualism."

Here the brain matters even more. Each moral system is, in the brain, a system of neural circuitry. How can inconsistent systems function smoothly in the same brain? The answer is twofold: (1) mutual inhibition (when one system is turned on the other is turned off); and (2) neural binding to different issues (when each system operates on different concerns).

Biconceptualism is central to our politics, and it is vital to understand how it works. We will be discussing it throughout this book.

What Is Rationality?

The brain and cognitive sciences have radically changed our understanding of what reason is and what it means to be rational. Unfortunately, all too many progressives have been taught a false and outdated theory of reason itself, one in which framing, metaphorical thought, and emotion play no role in rationality. This has led many progressives to the view that the facts—alone—will set you free. Progressives are constantly giving lists of facts.

Facts matter enormously, but to be meaningful they must be framed in terms of their moral importance. Remember, you can only understand what the frames in your brain allow you to understand. If the facts don't fit the frames in your brain, the frames in your brain stay and the facts are ignored or challenged or belittled. We will explore those frames in detail in the pages ahead.

It is by popular demand that this book is short and informal. It is meant to be a practical guide both for citizen activists and for anyone with a serious interest in politics. Those who want a more systematic and scholarly treatment should read my books *Moral Politics: How Liberals and Conservatives Think* (second edition), *Thinking Points, Whose Freedom?, The Political Mind,* and *The Little Blue Book* (with Elisabeth Wehling). And for those just dying to read clearly written 600-page academic books and hundreds of articles on both political and academic topics, you can find them on my website: www.georgelakoff.com. But for a quick informative read and your first introduction to framing, start here.

It is vital—for us, for our country, and for the world—that we understand the progressive values on which this country was founded and that made it a great democracy. If we are to keep that democracy, we must learn to articulate those values loud and clear. If progressives are to win in the future, we must present a clear moral vision to the country—a moral vision common to all progressives. It must be more than a laundry list of facts, policies, and programs. It must present a moral alternative, one traditionally American, one that lies behind everything Americans are proud of.

This update of the original version of *Don't Think of an Elephant!* is written in the service of that vision.

Enjoy!

★★ PART I ★★

FRAMING 101:
THEORY AND APPLICATION

Framing 101: How to Take Back Public Discourse

—January 21, 2004—

On this date I spoke extemporaneously to a group of about two hundred progressive citizen-activists in Sausalito, California. Some updates have been made.

When I teach the study of framing at Berkeley, in Cognitive Science 101, the first thing I do is I give my students an exercise. The exercise is: Don't think of an elephant! Whatever you do, do not think of an elephant. I've never found a student who is able to do this. Every word, like *elephant*, evokes a frame, which can be an image or other kinds of knowledge: Elephants are large, have floppy ears, tusks, and a trunk, live naturally in jungles, are associated with circuses, and so on. The word is defined relative to that frame. When we negate a frame, we evoke the frame.

Richard Nixon found that out the hard way. While under pressure to resign during the Watergate scandal, Nixon addressed the nation on TV. He stood before the nation and said, "I am not a crook." And everybody thought about him as a crook.

This gives us a basic principle of framing: When you are arguing against the other side, do not use their language. Their language picks out a frame—and it won't be the frame you want.

Let me give you an example. On the day that George W. Bush arrived in the White House, the phrase *tax relief* started coming out of the White House. It was repeated almost every day thereafter, was used by the press in describing his policies, and slowly became so much a part of public discourse that liberals started using it.

Think of the framing for *relief*. For there to be relief, there must be an affliction, an afflicted party, and a reliever who removes the affliction and is therefore a hero. And if people try to stop the hero, those people are villains for trying to prevent relief.

When the word *tax* is added to *relief,* the result is a metaphor: Taxation is an affliction. And the person who takes it away is a hero, and anyone who tries to stop him is a bad guy. This is a frame. It is made up of ideas, like *affliction* and *hero.* The language that evokes the frame comes out of the White House, and it goes into press releases, goes to every radio station, every TV station, every newspaper. And soon the *New York Times* is using *tax relief.* And it is not only on Fox; it is on CNN, it is on NBC, it is on every station because it is "the president's tax-relief plan." And soon the Democrats are using *tax relief*—and shooting themselves in the foot.

It is remarkable. We have seen Democrats adopting the conservative view of taxation as an affliction when they have offered "tax relief for the middle class."

They were accepting the conservative frame. The conservatives had set a trap: The words draw you into their worldview.

That is what framing is about. Framing is about getting language that fits your worldview. It is not just language. The ideas are primary—and the language carries those ideas, evokes those ideas.

There was another noteworthy example of conservative framing in George W. Bush's State of the Union address in January 2005. This one was a remarkable metaphor to find in a State of the Union address. President Bush said, "We do not need a permission slip to defend America." What was going on with a permission slip? He could have just said, "We won't ask permission." But talking about a permission slip is different. Think about when you last needed a permission slip. Think about who has to ask for a permission slip. Think about who is being asked. Think about the relationship between them.

Those are the kinds of questions you need to ask if you are to understand contemporary political discourse. While you are con-templating them, I want to raise other questions for you.

My work on politics began when I asked myself just such a ques-tion. It was back in the fall of 1994. I was watching election speeches and reading the Republicans' "Contract with America." The ques-tion I asked myself was this: What do the conservatives' positions on issues have to do with each other? If you are a conservative, what does your position on abortion have to do with your position on taxation? What does that have to do with your position on the environment? Or foreign policy? How do these positions fit together? What does being against gun control have to do with being for tort reform? What makes sense of the linkage? I could not figure it out. I said to myself, *These are strange people. Their collection of positions makes no sense.* But then a thought occurred to me. *I have exactly the opposite position on every issue. What do my positions have to do with one another?* And I could not figure that out either.

That was extremely embarrassing for someone who does cognitive science and linguistics.

Eventually the answer came. And it came from a very unexpected place. It came from the study of family values. I had asked myself why conservatives were talking so much about family values. And why did certain values count as "family values" while others did not? Why would anyone in a presidential campaign, in congressional campaigns, and so on, when the future of the world was being threatened by nuclear proliferation and global warming, constantly talk about family values?

At this point I remembered a paper that one of my students had written some years back that showed that we all have a metaphor for the nation as a family. We have Founding Fathers. The Daughters of the American Revolution. We "send our sons" to war. This is a natural metaphor because we usually understand large social groups, like nations, in terms of small ones, like families or communities.

Given the existence of the metaphor linking the nation to the family, I asked the next question: If there are two different understandings of the nation, do they come from two different understandings of family?

I worked backward. I took the various positions on the conservative side and on the progressive side and I said, "Let's put them through the metaphor from the opposite direction and see what comes out." I put in the two different views of the nation, and out popped two different models of the family: a strict father family and a nurturant parent family. You know which is which.

Now, when I first did this—and I'll tell you about the details in a minute—I was asked to give a talk at a linguistics convention. I decided I would talk about this discovery. In the audience were two members of the Christian Coalition who were linguists and good friends of mine. Excellent linguists. And very, very good people. Very nice people. People I liked a lot. They took me aside at the party afterward and said, "Well, this strict father model of the family, it's close, but not quite right. We'll help you get the details right. However, you should know all this. Have you read Dobson?"

I said, "Who?"

They said, "James Dobson."

I said, "Who?"

They said, "You're kidding. He's on three thousand radio stations."

I said, "Well, I don't think he's on NPR. I haven't heard of him."

They said, "Well, you live in Berkeley."

"Where would I . . . does he write stuff?"

"Oh," they said, "oh yes. He has sold millions of books. His classic is *Dare to Discipline*."

My friends were right. I followed their directions to my local Christian bookstore, and there I found it all laid out: the strict father model in all its details. Dobson at the time was an influential figure in conservative politics, with a 100-to-200-million-dollar-a-year operation, a widely distributed and read column in newspapers all over America, as well as his own zip code because so many people were writing to order his books and pamphlets. He was effectively teaching people how to use the strict father model to raise their kids, and he understood the connection between strict father families, right-wing politics, evangelical religion, laissez-faire economics, and neoconservative foreign policy.

The strict father model begins with a set of assumptions: The world is a dangerous place, and it always will be, because there is evil out there in the world. The world is also difficult because it is competitive. There will always be winners and losers. There is an absolute right and an absolute wrong. Children are born bad, in the sense that they just want to do what feels good, not what is right. Therefore, they have to be made good.

What is needed in this kind of a world is a strong, strict father who can:

- protect the family in the dangerous world,
- support the family in the difficult world, and
- teach his children right from wrong.

What is required of the child is obedience, because the strict father is a moral authority who knows right from wrong. It is further assumed that the only way to teach kids obedience—that is, right from wrong—is through punishment, painful punishment, when they do wrong.

This includes hitting them, and some authors on conservative child rearing recommend sticks, belts, and wooden paddles on the bare bottom. Some authors suggest this start at birth, but Dobson was more liberal. "There is no excuse for spanking babies younger than fifteen or eighteen months of age" (Dobson, *The New Dare to Discipline*, 65).

The rationale behind physical punishment is this: When children do something wrong, if they are physically disciplined, they learn not to do it again. That means that they will develop internal discipline to keep themselves from doing wrong, so that in the future they will be obedient and act morally. Without such punishment, the world will go to hell. There will be no morality.

Such internal discipline has a secondary effect. It is what is required for success in the difficult, competitive world. That is, if people are disciplined and pursue their self-interest in this land of opportunity, they will become prosperous and self-reliant. Thus, the strict father model links morality with prosperity. The same discipline you need to be moral is what allows you to prosper. The link is individual responsibility and the pursuit of self-interest. Given opportunity, individual responsibility, and discipline, pursuing your self-interest should enable you to prosper.

Now, Dobson was very clear about the connection between the strict father worldview and free market capitalism. The link is the morality of self-interest, which is the conservative version of Adam Smith's view of capitalism. Adam Smith said that if everyone pursues their own profit, then the profit of all will be maximized by the invisible hand—that is, by nature—just naturally. Go about pursuing your own profit, and you are helping everyone.

This is linked to a general metaphor that views well-being as wealth. For example, if I do you a favor, you say, "I owe you one," or "I'm in your debt." Doing something good for someone is metaphorically like giving him money. He "owes" you something. And he says, "How can I ever repay you?"

Apply this metaphor to Adam Smith's "law of nature": If everyone pursues her own self-interest, then by the invisible hand, by nature, the self-interest of all will be maximized. That is, it is moral to pursue your self-interest, and there is a name for those people who do not do it. The name is do-gooder. A do-gooder is someone who is trying to help someone else rather than herself and is getting in the way of those who are pursuing their self-interest. Do-gooders screw up the system.

In this model there is also a definition of what it means to become a good person. A good person—a moral person—is someone who is disciplined enough to be obedient to legitimate authority, to learn what is right, to do what is right and not do what is wrong, and to pursue her self-interest to prosper and become self-reliant. A good child grows up to be like that. A bad child is one who does not learn discipline, does not function morally, does not do what is right, and therefore is not disciplined enough to become prosperous. She cannot take care of herself and thus becomes dependent.

When the good children are mature, they either have learned discipline and can prosper, or have failed to learn it. From this point on the strict father is not to meddle in their lives.

This translates politically into no government meddling. Consider what all this means for social programs: It is immoral to give people things they have not earned, because then they will not develop discipline and will become both dependent and immoral. This theory says that social programs are immoral because they make people dependent. Promoting social programs is immoral. And what does this say about budgets? Well, if there are a lot of progressives in Congress who think that there should be social programs, and if you believe that social programs are immoral, how do you stop these immoral people?

In the strict father frame, it is quite simple. What you have to do is reward the good people—the ones whose prosperity reveals their discipline and hence their capacity for morality—with a tax cut, and make it big enough so that there is not enough money left for social programs. As Grover Norquist says, it "starves the beast."

For example, as a result of the Republican House's refusal to either cut tax loopholes or raise taxes to pay its bills, the 2013 "sequester"— an across-the-board cut in government programs—was put into place. Here are some examples of what was cut from various government agencies' budgets, from a February 20, 2013, article in the *Washington Post*:

- The National Institutes of Health: cut by $1.6 billion.
- The Centers for Disease Control and Prevention: cut by about $303 million.
- Head Start: cut by over $400 million, kicking 57,000 kids out of the program.
- FEMA's disaster relief budget: cut by $928 million.
- Public housing support: cut by about $1.74 billion.
- The FDA: cut by $209 million.
- NASA: cut by $896 million.
- Special education: cut by $827 million.
- The Energy Department's programs for securing our nukes: cut by $903 million.
- The National Science Foundation: cut by about $361 million.
- State Department diplomatic functions: cut by $665 million.
- Global health programs: cut by $411 million.
- The Nuclear Regulatory Commission: cut by $53 million.
- The SEC: cut by $74 million.
- The United States Holocaust Memorial Museum: cut by $3 million.
- The Library of Congress: cut by $30 million.
- The Patent and Trademark Office: cut by $148 million.

Conservatives see this as cutting "wasteful spending"—that is, spending for "bad" social programs.

Are conservatives against all government? No. They are not against the military; they are not against homeland security; they are not against tax cuts, loopholes, and subsidies for corporations;

they are not against the conservative Supreme Court. There are many aspects of government that they like very much. Subsidies for corporations, which reward the good people—the investors in those corporations—are great. No problem there.

But they are against nurturance and care. They are against social programs that take care of people—early childhood education, Medicaid for the poor, raising the minimum wage, unemployment insurance. That is what they see as wrong. That is what they are trying to eliminate on moral grounds. That is why they are not merely a bunch of crazies or mean and greedy—or stupid—people, as many liberals believe. What is even scarier is that conservatives are acting on principle, on what they believe is moral. And they have supporters around the country. People who have strict father morality and who apply it to politics are going to believe that this is the right way to govern.

Think for a minute about what this says about foreign policy. Suppose you are a moral authority. As a moral authority, how do you deal with your children? Do you ask them what they should do or what you should do? No. You tell them. What the father says, the child does. No back talk. Communication is one-way. It is the same with foreign policy. That is, the president does not engage in diplomacy or ask the help of allies; the president tells. If you are a moral authority, you know what is right, you have power, and you use it. You would be immoral yourself if you abandoned your moral authority.

Map this onto foreign policy, and it says that you cannot give up sovereignty. The United States, being the best and most powerful country in the world—a moral authority—should not be asking anybody else what to do. We should be using our military power.

This belief comes together with a set of metaphors that have run foreign policy for a long time. There is a common metaphor learned in graduate school classes on international relations. It is called the rational actor metaphor. It is the basis of classical "realist" international relations theory, and in turn it assumes another metaphor: that every nation is a person. Therefore there are "rogue states," there are "friendly nations," and so on. And there is a national interest.

What does it mean, in this worldview, to act in your self-interest? In the most basic sense it means that you act in ways that will help you be healthy and strong. In the same way, by the metaphor that a nation is a person, it is good for a nation to be healthy (that is, economically healthy—defined as having a large GDP) and strong (that is, militarily strong). It is not necessary that all the individuals in the country be healthy, but the companies should be, and the country as a whole should have a lot of money. That is the idea.

The question is: How do you maximize your self-interest? That is what foreign policy is about: maximizing self-interest—not working for the interest of all. The rational actor metaphor says that every actor, every person, is rational, and that it is irrational to act against your self-interest. Therefore it is rational for every person to act to maximize self-interest. Then by the further metaphor that nations are persons ("friendly nations," "rogue states," "enemy nations," and so on), there are adult nations and child nations, where adulthood is industrialization. The child nations are called "developing" nations or "underdeveloped" states. Those, again in this view, are the backward ones. And what should we do? If you are a strict father, you tell the children how to develop, tell them what rules they should follow, and punish them when they do wrong. That is, you operate using, say, the policies of the International Monetary Fund.

And who is in the United Nations? Most of the United Nations consists of developing and underdeveloped countries. That means they are metaphorical children. Now let's go back to the State of the Union address. Should the United States have consulted the United Nations and gotten its permission to invade Iraq? An adult does not "ask for a permission slip"! The phrase itself, permission slip, puts you back in grammar school or high school, where you need a permission slip from an adult to go to the bathroom. You do not need to ask for a permission slip if you are the teacher, if you are the principal, if you are the person in power, the moral authority. The others should be asking you for permission. That is what the permission slip phrase in the 2004 State of the Union address was about. Every conservative in the audience got it. They got it right away.

Two powerful words: permission slip. What Bush did was evoke the adult–child metaphor for other nations. He said, "We're the adult in charge." He was operating in the strict father worldview, and it did not have to be explained. It is evoked automatically. This is what is done regularly by the conservatives.

Finally, there is the conservative view of the moral hierarchy. As we have seen, the rich and those who can take care of themselves are considered more moral than the poor and those who need help. But moral superiority on a wider scope is central to conservative thought. The basic idea is that those who are more moral should rule. How do you know who is more moral? Well, in a well-ordered world (ordered by God), the moral have come out on top. Here is the hierarchy: God above man; man above nature; adults above children; Western culture above non-Western culture; our country above other countries. These are general conservative values. But the hierarchy goes on, and it explains the oppressive views of more radical conservatives: men above women, Christians above non-Christians, whites above nonwhites, straights above gays.

Thus, disobedient children in southern states can be "paddled" in school with sticks by teachers; women seeking abortions must undergo embarrassing medical procedures, and notification of husbands and fathers; African Americans and Hispanics have voting rights taken away; legislation against gay marriage is passed by conservative legislatures. In short, the moral hierarchy is an implicit part of the culture wars.

Now let me talk a bit about how progressives understand their morality and what their moral system is. It too comes out of a family model, what I call the nurturant parent model. The strict father worldview is so named because according to its own beliefs, the father is the head of the family. The nurturant parent worldview is gender neutral.

Both parents are equally responsible for raising the children. The assumption is that children are born good and can be made better. The world can be made a better place, and our job is to work on that. The parents' job is to nurture their children and to raise their children to be nurturers of others.

What does nurturance mean? It means three things: empathy, responsibility for yourself and others, and a commitment to do your best not just for yourself, but for your family, your community, your country, and the world. If you have a child, you have to know what every cry means. You have to know when the child is hungry, when she needs a diaper change, when she is having nightmares. And you have a responsibility—you have to take care of the child. Since you cannot take care of someone else if you are not taking care of yourself, you have to take care of yourself enough to be able to take care of the child.

All this is not easy. Anyone who has ever raised a child knows that it is hard. You have to be strong. You have to work at it. You have to be very competent. You have to know a lot.

In addition, all sorts of other values immediately follow from empathy, responsibility for yourself and others, and commitment to do your best for all. Think about it.

First, if you empathize with your child, you will provide protection. This comes into politics in many ways. What do you protect your child from? Crime and drugs, certainly. You also protect your child from cars without seat belts, from smoking, from poisonous additives in food. So progressive politics focuses on environmental protection, worker protection, consumer protection, and protection from disease. These are the things that progressives want the government to protect their citizens from. But there are also terrorist attacks, which liberals and progressives have not been very good at talking about in terms of protection. Protection is part of the progressive moral system, but it has not been elaborated on enough. And on September 11, 2001, progressives did not have a whole lot to say. That was unfortunate, because nurturant parents and progressives do care about protection. Protection is important. It is part of our moral system.

Second, if you empathize with your child, you want your child to be fulfilled in life, to be a happy person. And if you are an unhappy, unfulfilled person yourself, you are not going to want other people to be happier than you are. The Dalai Lama teaches us that. Therefore it is your moral responsibility to be a happy, fulfilled person. Your moral responsibility! Further, it is your moral responsibility to teach your child to be a happy, fulfilled person who wants others to be happy and fulfilled. That is part of what nurturing family life is about. It is a common precondition for caring about others.

There are still other nurturant values.

- If you want your child to be fulfilled in life, the child has to be free enough to seek and possibly find fulfillment. Therefore **freedom** is a value.
- You do not have very much freedom if there is no opportunity or prosperity. Therefore **opportunity** and **prosperity** are progressive values.
- If you really care about your child, you want your child to be treated fairly by you and by others. Therefore **fairness** is a value.
- If you are connecting with your child and you empathize with that child, you have to have open, two-way communication. Honest, open communication. That becomes a value.
- You live in a community, and that community will affect how your child grows up. Therefore **community-building, service to the community**, and **cooperation in a community** become values.
- To have cooperation, you must have **trust**, and to have trust, you must have **honesty** and open two-way communication. Trust, honesty, and open communication are fundamental progressive values—in a community as in a family.

These are the nurturant values—and they are the progressive values. As a progressive, you have them. You know you have them. You recognize them.

Every progressive political program is based on one or more of these values. That is what it means to be a progressive.

There are several types of progressives. How many types? I am asking as a cognitive scientist, not as a sociologist or a political scientist. From the point of view of a cognitive scientist, who looks at modes of thought, there are six basic types of progressives, each with a distinct mode of thought. They share all the progressive values, but are distinguished by some differences.

- **Socioeconomic progressives** think that everything is a matter of money and class and that all solutions are ultimately economic and social class solutions.

- **Identity politics progressives** say it is time for their oppressed group to get its share now.
- **Environmentalists** think in terms of sustainability of the earth, the sacredness of the earth, and the protection of native peoples. And they recognize that global warming is the major moral challenge of our time, making all other issues pale by comparison.
- **Civil liberties progressives** want to maintain freedoms against threats to freedom.
- **Spiritual progressives** have a nurturant form of religion or spirituality. Their spiritual experience has to do with their connection to other people and the world, and their spiritual practice has to do with service to other people and to their community. Spiritual progressives span the full range from Catholics and Protestants to Jews, Muslims, Buddhists, Goddess worshippers, and pagan members of Wicca.
- **Antiauthoritarians** say there are all sorts of illegitimate forms of authority out there and we have to fight them, whether they are big corporations or anyone else.

All six types are examples of nurturant parent morality. The problem is that many of the people who have one of these modes of thought do not recognize that theirs is just one special case of something more general, and do not see the unity in all the types of progressives. They often think that theirs is the only way to be a true progressive. That is sad. It keeps people who share progressive values from coming together. We have to get past that harmful idea. The other side did. Until the Tea Party came along.

Back in the 1950s conservatives hated each other. The financial conservatives hated the social conservatives. The libertarians did not get along with the social conservatives or the religious conservatives. And many social conservatives were not religious. A group of conservative leaders got together around William F. Buckley Jr. and others and started asking what the different groups of conservatives had in common and whether they could agree to disagree in order to promote a general conservative cause. They started magazines and think tanks, and invested billions of dollars. The first thing they did, their first victory, was getting Barry Goldwater nominated in 1964. He lost, but when he lost, they went back to the drawing board and put more money into organization.

During the Vietnam War, they noticed that most of the bright young people in the country were not becoming conservatives. Conservative was a dirty word. Therefore, in 1970, Lewis Powell, just two months before he became a Supreme Court justice appointed by Nixon (at the time he was the chief counsel to the US Chamber of Commerce), wrote a memo—the Powell memo. It was a fateful document. He said that the conservatives had to keep the country's best and brightest young people from becoming antibusiness. What we need to do, Powell said, is set up institutes within the universities and outside the universities. We have to do research, we have to write books, we have to endow professorships to teach these people the right way to think.

After Powell went to the Supreme Court, these ideas were taken up by William Simon, secretary of the treasury under Nixon. He convinced some very wealthy people and families with foundations—Coors, Scaife, Olin—to set up the Heritage Foundation, the Olin professorships, the Olin Institute at Harvard, and other institutions. These institutes have done their job very well. People associated with them have written more books than the people on the left have, on all issues. The conservatives support their intellectuals. They create media opportunities. They have media studios down the hall in their institutes so that getting on TV is easy.

When the amount of research money spent by the right over a period of time is compared with the amount of media time during that period, we see a direct correlation. At present, the Koch brothers are pouring money into right-wing campaigns.

This is not an accident. Conservatives, through their think tanks, figured out the importance of framing, and they figured out how to frame every issue. They figured out how to get those frames out there, how to get their people in the media all the time. They set up training institutes. The Leadership Institute in Virginia trains tens of thousands of conservatives a year and runs constant programs around the United States and in fifteen foreign countries. Trained conservative spokespeople receive regular talking points and are booked by booking agencies on radio, TV, and other local venues.

Conservatives figured out how to bring their people together. Every Wednesday, Grover Norquist has a group meeting—around eighty people—of leaders from the full range of the right. They are invited, and they debate. They work out their differences, agree to disagree, and when they disagree, they trade off. The idea is: This week he'll win on his issue. Next week, I'll win on mine. Each one may not get everything he wants, but over the long haul, he gets a lot of what he wants. The meetings have gone on for two decades. In recent years, the Wednesday morning Norquist meetings have expanded to forty-eight states. Via ALEC (American Legislative Exchange Council), conservatism has spread at the state level, allowing conservatives to take over state legislatures, gerrymander congressional districts, and take over the House of Representatives with a minority of national voter support.

It is only in the wake of the 2008 Obama sweep that the radical conservative Tea Party movement has split from the previously unified conservative movement.

The progressive world has not caught up.

And what is worse is a set of myths believed by liberals and progressives. These myths come from a good source, but they end up hurting us badly.

The myths began with the Enlightenment, and the first one goes like this:

The truth will set us free. If we just tell people the facts, since people are basically rational beings, they'll all reach the right conclusions.

But we know from cognitive science that people do not think like that. People think in frames. The strict father and nurturant parent frames each force a certain logic. To be accepted, the truth must fit people's frames. If the facts do not fit a frame, the frame stays and the facts bounce off. Why?

Neuroscience tells us that each of the concepts we have—the long-term concepts that structure how we think—is instantiated in the synapses of our brains. Concepts are not things that can be changed just by someone telling us a fact. We may be presented with facts, but for us to make sense of them, they have to fit what is already in the synapses of the brain. Otherwise facts go in and then they go right back out. They are not heard, or they are not accepted as facts, or they mystify us: Why would anyone have said that? Then we label the fact as irrational, crazy, or stupid. That's what happens when progressives just "confront conservatives with the facts." It has little or no effect, unless the conservatives have a frame that makes sense of the facts.

Similarly, a lot of progressives hear conservatives talk and do not understand them because they do not have the conservatives' frames. They assume that conservatives are stupid.

They are not stupid. They are winning because they are smart. They understand how people think and how people talk. They think! That is what those think tanks are about. They support their intellectuals. They write all those books. They put their ideas out in public.

There are certainly cases where conservatives have lied. That is true. Of course, it is not true that only conservatives lie. But it is true that there were significant lies—even daily lies—by the Bush administration.

However, it is equally important to recognize that many of the ideas that outrage progressives are what conservatives see as truths—presented from their point of view. We must distinguish cases of out-and-out distortion, lying, and so on, from cases where conservatives are presenting what they consider truth.

Is it useful to go and tell everyone what the lies are? Well, it is certainly not useless or harmful for us to know when they are lying. But also remember that the truth alone will not set you free.

The scientific facts about global warming are stated and restated day after day around the country, but they fall on conservative deaf brains—brains with frames that don't fit those facts.

There is another myth that also comes from the Enlightenment, and it goes like this: It is irrational to go against your self-interest, and therefore a normal person, who is rational, reasons on the basis of self-interest. Modern economic theory and foreign policy are set up on the basis of that assumption.

The myth has been challenged by cognitive scientists such as Daniel Kahneman (who won the Nobel Prize in economics for his theory) and Amos Tversky, who have shown that people do not really think that way. Nevertheless, most of economics is still based on the assumption that people will naturally always think in terms of their self-interest.

This view of rationality comes into Democratic politics in a very important way. It is assumed that voters will vote their self-interest. Democrats are shocked or puzzled when voters do not vote their self-interest. "How," Democrats keep asking me, "can any poor person vote for Republicans when Republican policies hurt them so badly?" The Democratic response is to try to explain over and over to the conservative poor why voting Democratic would serve their self-interest. Despite all evidence that this is a bad strategy, Democrats keep banging their heads against the wall.

In the 2012 election, Democrats argued that Mitt Romney's policies would only help the rich. But most poor conservatives still voted Republican against their self-interest, even though Romney was recorded saying not very nice things about the poor in general.

It is claimed that about a third of the populace thinks that they are, or someday will be, in the top 1 percent, and that for this reason they vote on the basis of a hoped-for future self-interest. But what about the other two-thirds, who have no dream that they will ever get super-rich? They are clearly not voting in their self-interest, or even their hoped-for future self-interest.

People do not necessarily vote in their self-interest. They vote their identity. They vote their values. They vote for who they identify with. They may identify with their self-interest. That can happen. It is not that people never care about their self-interest. But they vote their identity. And if their identity fits their self-interest, they will vote for that. It is important to understand this point. It is a serious mistake to assume that people are simply always voting in their self-interest.

A third mistake is this: There is a metaphor that political campaigns are marketing campaigns where the candidate is the product and the candidate's positions on issues are the features and qualities of the product. This leads to the conclusion that polling should determine which issues a candidate should run on. Which issue shows the highest degree of support for a candidate's position? If it's prescription drugs, you run on a platform featuring prescription drugs. Is it keeping social security? Then you run on a platform featuring social security. You make a list of the top issues, and those are the issues you run on. You also do market segmentation: District by district, you find out the most important issues, and those are the ones you talk about when you go to that district.

It does not work. Sometimes it can be useful, and, in fact, the Republicans use it in addition to their real practice. But their real practice, and the real reason for their success, is this: They say what they idealistically believe. They say it; they talk to their base using the frames of their base. Liberal and progressive candidates tend to follow their polls and decide that they have to become more "centrist" by moving to the right. The conservatives do not move at all to the left, and yet they win!

Why? What is the electorate like from a cognitive point of view? Probably 35 to 40 percent of people have a strict father model governing their politics. Similarly, there are people who have a nurturant view governing their politics, probably another 35 to 40 percent. And then there are all the people who are said to be in the "middle."

There is no ideology of the middle. There is no moral system or political position that defines the "middle." The people in the "middle" are largely biconceptuals, people who are conservative on some issues and progressive on others, in all sorts of combinations.

Notice that I said *governing* their politics. We all have both models, either actively or passively. Progressives see a John Wayne movie or an Arnold Schwarzenegger movie, and they can understand it. They do not say, "I don't know what's going on in this movie." They have a strict father model, at least passively. And if you are a conservative and you understand Oprah, you have a nurturant parent model, at least passively. Everyone has both worldviews because both worldviews are widely present in our culture, but people do not necessarily live by one worldview all of the time.

So the question is: Are you living by one of the family-based models? But that question is not specific enough. There are many aspects of life, and many people live by one family-based model in one part of their lives and another in another part of their lives. I have colleagues who are nurturant parents at home and liberals in their politics, but strict fathers in their classrooms. Reagan knew that blue-collar workers who were nurturant in their union politics were often strict fathers at home. He used political metaphors that were based on the home and family, and got them to extend their strict father way of thinking from the home to politics.

This is very important to understand. The goal is to activate your model in the people in the "middle." The people who are in the middle have both models, used regularly in different parts of their lives. What you want to do is to get them to use your model for politics—to activate your worldview and moral system in their political decisions. You do that by talking to people using frames based on your worldview.

However, in doing that, you do not want to offend the people in the middle who have up to this point made the opposite choice. Since they have and use both models in their lives, they might still be persuaded to activate the opposite model for politics.

Clinton figured out how to handle this problem. He stole the other side's language. He talked about "welfare reform," for example. He said, "The age of big government is over." He did what he wanted to do, only he took their language and used their words to describe it. It made them very mad.

It turns out that what is good for the goose is good for the gander, and guess what? When George W. Bush arrived, we got "compassionate conservatism." The Clear Skies Initiative. Healthy Forests. No Child Left Behind. This is the use of language to mollify people who have nurturant values, while the real policies are strict father policies. This can even attract the people in the middle who might have qualms about you. This is the use of Orwellian language—language that means the opposite of what it says—to appease people in the middle at the same time as you pump up the base. That is part of the conservative strategy.

Liberals and progressives typically react to this strategy in a self-defeating way. The usual reaction is, "Those conservatives are bad people; they are using Orwellian language. They are saying the opposite of what they mean. They are deceivers. Bad. Bad."

All true. But we should recognize that they use Orwellian language precisely when they have to: when they are weak, when they cannot just come out and say what they mean. Imagine if they came out supporting a "Dirty Skies Bill" or a "Forest Destruction Bill" or a "Kill Public Education" bill. They would lose. They are aware people do not support what they are really trying to do.

Orwellian language points to weakness—Orwellian weakness. When you hear Orwellian language, note where it is, because it is a guide to where they are vulnerable. They do not use it everywhere. It is very important to notice this and use their weakness to your advantage.

A very good example relates to the environment. The right's language man is Frank Luntz, who puts out books of language guidelines, which are used as training manuals for conservative candidates, as well as lawyers, judges, and other public speakers—even high school students who want to be conservative public figures. In these books, Luntz tells you what language to use for a conservative advantage.

It was Luntz who persuaded conservatives to stop talking about "global warming" because it sounded too scary and suggested human agency. Instead, he brought "climate change" into our public discourse on the grounds that "climate" sounded kind of nice (think palm trees) and change just happens, with no human agency. By 2003, with the scientific consensus going against conservatives, Luntz suggested Orwellian language. He suggested using words like *healthy*, *clean*, and *safe* even when talking about coal or nuclear power plants. Hence "clean coal." Conservative legislation that increases pollution is called the Clear Skies Act. He is supporting global warming denial by suggesting that people say that the science is not settled and that our economy should not be threatened. Recently, his focus group research showed support for cap and trade legislation. He has suggested using the language of "energy independence," which supports continued fracking, but not talking about saving the planet.

Luntz once wrote a memo for talking to women. How do you talk to women? According to Luntz, women like certain words, so when you are talking to an audience of women, here are the words you use as many times as possible: *love, from the heart,* and *for the children.* And if you read George W. Bush's speeches from that period, *love, from the heart,* and *for the children* show up over and over again.

This kind of language use is a science. Like any science, it can be used honestly or harmfully. This kind of language use is taught. This kind of language use is also a discipline. Conservatives enforce message discipline. In many offices there is a pizza fund: Every time you use the "wrong" language, you have to put a quarter in the pizza fund. People quickly learn to say *tax relief* or *partial-birth abortion,* not something else.

But Luntz is about much more than language. He recognizes that the right use of language starts with ideas—with the right framing of the issues, a framing that reflects a consistent conservative moral perspective, what we have called strict father morality. Luntz's writing is not just about language. For each issue, he explains what the conservative reasoning is, what the progressive reasoning is, and how the progressive arguments can be best attacked from a conservative perspective. He is clear: Ideas come first.

One of the major mistakes liberals make is that they think they have all the ideas they need. They think that all they lack is media access. Or maybe some magic bullet phrases, the liberal equivalent of *partial-birth abortion.*

When you think you just lack words, what you really lack are ideas. Ideas come in the form of frames. When the frames are there, the words come readily. There's a way you can tell when you lack the right frames. There's a phenomenon you have probably noticed. A conservative on TV uses two words, like *tax relief.* And the progressive has to go into a paragraph-long discussion of his own view. The conservative can appeal to an established frame, that taxation is an affliction or burden, which allows for the two-word phrase *tax relief.* But there is no established frame on the other side. You can talk about it, but it takes some doing because there is no established frame, no fixed idea already out there.

In cognitive science there is a name for this phenomenon. It's called hypocognition—the lack of the ideas you need, the lack of a relatively simple fixed frame that can be evoked by a word or two.

The idea of hypocognition comes from a study in Tahiti in the 1950s by the late anthropologist Bob Levy, who was also a therapist. Levy addressed the question of why there were so many suicides in Tahiti, and discovered that Tahitians did not have a concept of grief. They felt grief. They experienced it. But they did not have a concept for it or a name for it. They did not see it as a normal emotion. There were no rituals around grief. No grief counseling, nothing like it. They lacked a concept they needed—and wound up committing suicide all too often.

Progressives are suffering from massive hypocognition. The conservatives used to suffer from it. When Goldwater lost in 1964, they had very few of the concepts that they have today. In the intermediate fifty years, conservative thinkers have filled in their conceptual gaps. But our conceptual gaps are still there.

Let's go back to tax relief.

What is taxation? Taxation is what you pay to live in a civilized country—what you pay to have democracy and opportunity, and what you pay to use the infrastructure paid for by previous taxpayers: the highway system, the Internet, the entire scientific establishment, the medical establishment, the communications system, the airline system. All are or were paid for by taxpayers.

You can think of taxation metaphorically in at least two ways. First, as an investment. Imagine the following ad:

> Our parents invested in the future, ours as well as theirs, through their taxes. They invested their tax money in the interstate highway system, the Internet, the scientific and medical establishments, our communications system, our airline system, the space program. They invested in the future, and we are reaping the tax benefits, the benefits from the taxes they paid. Today we have assets—highways, schools and colleges, the Internet, airlines—that come from the wise investments they made.

Imagine versions of this ad running over and over, for years. Eventually, the frame would be established: Taxes are wise investments in the future.

Or take another metaphor:

> Taxation is paying your dues, paying your membership
> fee in America. If you join a country club or a commu-
> nity center, you pay fees. Why? You did not build the
> swimming pool. You have to maintain it. You did not
> build the basketball court. Someone has to clean it. You
> may not use the squash court, but you still have to pay
> your dues. Otherwise it won't be maintained and will
> fall apart. People who avoid taxes, like corporations that
> move to Bermuda, are not paying their dues to their
> country. It is patriotic to be a taxpayer. It is traitorous to
> desert our country and not pay your dues.

Perhaps Bill Gates Sr. said it best. In arguing to keep the inheritance tax, he pointed out that he and Bill Jr. did not invent the
Internet. They just used it—to make billions. There is no such thing
as a self-made man. Every businessman has used the vast American
infrastructure, which the taxpayers paid for, to make his money. He
did not make his money alone. He used taxpayer infrastructure. He
got rich on what other taxpayers had paid for: the banking system,
the Federal Reserve, the Treasury and Commerce Departments, and
the judicial system, where nine-tenths of cases involve corporate law.
These taxpayer investments support companies and wealthy investors. There are no self-made men! The wealthy have gotten rich using
what previous taxpayers have paid for. They owe the taxpayers of this
country a great deal and should be paying it back.

These are accurate views of taxes, but they are not yet enshrined
in our brains. They need to be repeated over and over again, and
refined until they take their rightful place in our synapses. But that
takes time. It does not happen overnight. Start now.

It is not an accident that conservatives are winning where they have
successfully framed the issues. They've got a forty- to fifty-year head
start. And more than two billion dollars in think tank investments.

And they are still thinking ahead. Progressives are not. Progressives feel so assaulted by conservatives that they can only think about

immediate defense. Democratic office holders are constantly under attack. Every day they have to respond to conservative initiatives. It is always, "What do we have to do to fight them off today?" This leads to politics that are reactive, not proactive.

And it is not just public officials. I have been talking to advocacy groups around the country, working with them and trying to help them with framing issues. I have worked with more than four hundred advocacy groups in this way. They have the same problems: They are under attack all the time, and they are trying to defend themselves against the next attack. Realistically, they do not have time to plan. They do not have time to think long-term. They do not have time to think beyond their particular issues.

They are all good people—intelligent, committed people. But they are constantly on the defensive. Why? It is not hard to explain it when we think about funding.

The right-wing think tanks get large block grants and endowments. Millions at a time. They are very well funded.

Furthermore, they know that they are going to get the money the next year, and the year after that. Remember, these are block grants— no strings attached. Do what you need. Hire intellectuals. Bring talent along. These institutions also build human capital for the future.

Progressive foundations spread the money around—thinly. They give twenty-five thousand dollars here, maybe fifty thousand, maybe even a hundred thousand. Sometimes it is a big grant. But recipients have to do something different from what everyone else is doing because the foundations see duplication as a waste of money. Not only that, but also they are not block grants like conservative foundations get; the recipients do not have full freedom to decide how to spend the money. And it is certainly not appropriate to use it for career development or infrastructure building or hiring intellectuals to think about long-term as well as short-term or interrelated policies. The emphasis is on providing direct services to the people who need the services: grassroots funding, not infrastructure creation. This is, for the most part, how progressive foundations work. And because of that, the organizations they fund have to have a very narrow focus. They have to have projects, not just areas they work on. Activists and advocates are overworked and underpaid, and they do not have time or energy to think about how they should be linking up with other people. They mainly do not have the time or training to think about framing their issues. The system forces a narrow focus—and with it, isolation.

You ask, "Why is it like this?" There is a reason. There is a deep reason, and it is a reason you should think about. In the right's hierarchy of moral values, the top value is preserving and defending the moral system itself. If that is your main goal, what do you do? You build infrastructure. You buy up media in advance. You plan ahead. You do things like give fellowships to right-wing law students to get them through law school if they join the Federalist Society. And you get them nice jobs after that. If you want to extend your worldview, it is very smart to make sure that over the long haul you have the people and the resources that you need.

On the left, the highest value is helping individuals who need help. So if you are a foundation or you are setting up a foundation, what makes you a good person? You help as many people as you can. And the more public budgets get cut, the more people there are who need help. So you spread the money around to the grassroots organizations, and therefore you do not have any money left for infrastructure or talent development, and certainly not for intellectuals. Do not waste a penny in duplicating efforts, because you have to help more and more people. How do you show that you are a good, moral person or foundation? By listing all the people you help; the more the better.

And so you perpetuate a system that helps the right. In the process, it also does help people. Certainly, it is not that people do not need help. They do. But what has happened as budgets and taxes get cut is that the right is privatizing the left. The right is forcing the left to spend ever more private money on what the government should be supporting.

There are many things that we can do about all this. Let's talk about where to start.

The right knows how to talk about values. We need to talk values. If we think about it a little, we can list our values. Bu not easy to think about how the values fit the issues, to know how to talk about every issue from the perspective of our values, not theirs.

Progressives also have to look at the integration of issues. This is something that the right is very, very savvy about. They know about what I call strategic initiatives. A strategic initiative is a plan in which a change in one carefully chosen issue area has automatic effects over many, many, many other issue areas.

For example, tax cuts. This seems straightforward, but as a result of tax cuts there is not enough money in the budget for any of the government's social programs. Not just not enough money for, say, homelessness or schools or environmental protection; instead, not enough money for everything at once, the whole range. This is a strategic initiative.

Or tort reform, which means putting limits on awards in lawsuits. Tort reform is a top priority for conservatives. Why do conservatives care so much about this? Well, as soon as you see the effects, you can see why they care. Because in one stroke you prohibit all of the potential lawsuits that will be the basis of future environmental legislation and regulation. That is, it is not just regulation of the chemical industry or the coal industry or the nuclear power industry or other things that are at stake. It is the regulation of everything. If parties who are harmed cannot sue immoral or negligent corporations or professionals for significant sums, the companies are free to harm the public in unlimited ways in the course of making money. And lawyers, who take risks and make significant investments in such cases, will no longer make enough money to support the risk. And corporations will be free to ignore the public good. That is what "tort reform" is about.

In addition, if you look at where Democrats get much of their money in the individual states, it is significantly from the lawyers who win tort cases. Many tort lawyers are important Democratic donors. Tort "reform"—as conservatives call it—cuts off this source of money. All of a sudden three-quarters of the money going to the Texas Democratic Party is not there. In addition, companies who poison the environment want to be able to cap possible awards. That way they can calculate in advance the cost of paying victims and build it into the cost of doing business. Irresponsible corporations win big from tort reform. The Republican Party wins big from tort reform. And these real purposes are hidden. The issue appears to be eliminating "frivolous lawsuits"—people getting thirty million dollars for having hot coffee spilled on them.

However, what the conservatives are really trying to achieve is not in the proposal. What they are trying to achieve follows from enacting the proposal. They don't care primarily about the lawsuits themselves. They care about getting rid of environmental, consumer, and worker protections in general. And they care about defunding the Democratic Party. That is what a strategic initiative is.

There have been a couple of strategic initiatives on the left—environmental impact reports and the Endangered Species Act—but it has been forty years since they were enacted.

Unlike the right, the left does not think strategically. We think issue by issue. We generally do not try to figure out what minimal change we can enact that will have effects across many issues. There are very few exceptions.

There are also strategic initiatives of another kind—what I call slippery slope initiatives: Take the first step and you're on your way off the cliff. Conservatives are very good at slippery slope initiatives. Take "partial-birth abortion." There are almost no such cases. Why do conservatives care so much? Because it is a first step down a slippery slope to ending all abortion. It puts out there a frame of abortion as a horrendous procedure, when most operations ending pregnancy are nothing like this.

Why an education bill about school testing? Once the testing frame applies not just to students but also to schools, then schools can, metaphorically, fail—and be punished for failing by having their allowance cut. Less funding in turn makes it harder for the schools to improve, which leads to a cycle of failure and ultimately elimination for many public schools. What replaces the public school system is a voucher system to support private schools. The wealthy would have good schools—paid for in part by what used to be tax payments for public schools. The poor would not have the money for good schools. We would wind up with a two-tier school system, a good one for the "deserving rich" and a bad one for the "undeserving poor."

The conservatives don't have to win on issue after issue after issue. There are many things a progressive can do about it. Here are eleven.

First, notice what conservatives have done right and where progressives have missed the boat. It is more than just control of the media, though that is far from trivial. What they have done right is to successfully frame the issues from their perspective. Acknowledge their successes and our failures.

Second, remember "Don't think of an elephant." If you keep their language and their framing and just argue against it, you lose because you are reinforcing their frame.

Third, the truth alone will not set you free. Just speaking truth to power doesn't work. You need to frame the truths effectively from your perspective.

Fourth, you need to speak from your moral perspective at all times. Progressive policies follow from progressive values. Get clear on your values and use the language of values. Drop the language of policy wonks.

Fifth, understand where conservatives are coming from. Get their strict father morality and its consequences clear. Know what you are arguing against. Be able to explain why they believe what they believe. Try to predict what they will say.

Sixth, think strategically, across issue areas. Think in terms of large moral goals, not in terms of programs for their own sake.

Seventh, think about the consequences of proposals. Form progressive slippery slope initiatives.

Eighth, remember that voters vote their identity and their values, which need not coincide with their self-interest.

Ninth, unite! And cooperate! Here's how: Remember the six modes of progressive thought: (1) socioeconomic, (2) identity politics, (3) environmentalist, (4) civil libertarian, (5) spiritual, and (6) antiauthoritarian. Notice which of these modes of thought you use most often—where you fall on the spectrum and where the people you talk to fall on the spectrum. Then rise above your own mode of thought and start thinking and talking from shared progressive values.

Tenth, be proactive, not reactive. Play offense, not defense. Practice reframing, every day, on every issue. Don't just say what you believe. Use your frames, not their frames. Use them because they fit the values you believe in.

Eleventh, speak to the progressive base in order to activate the nurturant model of biconceptual voters. Don't move to the right. Rightward movement hurts in two ways. It alienates the progressive base and it helps conservatives by activating their model in biconceptual voters.

★ ★ **PART II** ★ ★

FRAMING 102:
FRAMING THE UNFRAMED

2

Framing the Unframed

There are two common mistakes people make when thinking about framing.

The first mistake is believing that framing is a matter of coming up with clever slogans, like "death tax" or "partial-birth abortion," that resonate with a significant segment of the population. Those slogans only work when there has been a long—often decades-long—campaign of framing issues like taxation and abortion conceptually, so that the brains of many people are prepared to accept those phrases. I was once asked if I could reframe—that is, provide a winning slogan for—a global warming bill "by next Tuesday." I laughed. Effective reframing is the changing of millions of brains to be prepared to recognize a reality. That preparation hadn't been done.

The second mistake is believing that, if only we could present the facts about a certain reality in some effective way, then people would "wake up" to that reality, change their personal opinion, and start acting politically to change society. "Why can't people wake up?" is the complaint—as if people are "asleep" and just have to be aroused to see and comprehend the world around them. But the reality is that certain ideas have to be ingrained in us—developed over time consistently and precisely enough to create an accurate frame for our understanding.

Here is an example. Pensions, even by those who advocate for them, are often framed as benefits—"extras" granted by an employer to the employed. Yet what is a pension, really? A pension is delayed payment for work already done. As a condition for taking a job, a pension is part of your earned salary, withheld and invested by your employer, to be paid later, after retirement. So if an employer says, "we just don't have the money to pay for your pension," that means that he has either embezzled, stolen, or misspent your earnings, which by contract he is responsible for paying you. Your employer is a thief.

I've had the repeated experience of talking to union leaders and groups of workers, pointing out to them that a pension is delayed payment for work already done. I get universal agreement. Then I ask, "Have you ever said it?" "No." "Do you believe it?" "Yes." "Would you start saying it?" That is where it gets difficult. Even for progressives, it is hard to shake the frame constructed over years by pundits on the right that pensions are pay for not working.

Yet the fact that pensions are delayed payments is an obvious truth that would undermine the idea promulgated by pundits on the right that pensions are pay for not working.

So why can people perceive an important truth on a topic crucial to them, a truth that needs to be out in public, and not say it, not make it part of their everyday discourse?

The reason is that just telling someone something usually does not make it a neural circuit that they use every day or even a neural circuit that fits easily into their pre-existing brain circuitry—the neural circuits that define their previous understandings and forms of discourse.

It is difficult to say things that you are not sure the public is ready to hear, to say things that have not been said hundreds of times before.

As noted in chapter 1, this problem has a name—hypocognition—the lack of the overall neural circuitry that makes common sense of the idea and that fits the forms of communication that one normally engages in, the things you are ready to say and that the people you speak to are ready to hear.

Slogans can't overcome hypocognition. Only sustained public discussion has a chance. And that takes knowledge of the problem and a large-scale serious commitment to work for a change.

Several important issues that confront us right now—from global warming to the wealth gap and beyond—demand this kind of sustained discussion and commitment. I am offering this section of the book in the hope that various readers will take on the various tasks of working to provide frames—that is, automatic, effortless, everyday modes of understanding that we desperately need.

Reflexivity: The Brain and the World

You might think that the world exists independently of how we understand it. You would be mistaken.

Our understanding of the world is part of the world—a physical part of the world. Our conceptual framings exist in physical neural circuitry in our brains, largely below the level of conscious awareness, and they define and limit how we understand the world, and so they affect our actions in the world. The world is thus, in many ways, a reflection of how we frame it and act on those frames, creating a world in significant part framed by our actions. Accordingly, the frame-inherent world, structured by our framed actions, reinforces those frames and recreates those frames in others as they are born, grow, and mature in such a world.

This phenomenon is called reflexivity. The world reflects our understandings through our actions, and our understandings reflect the world shaped by the frame-informed actions of ourselves and others.

To function effectively in the world it helps to be aware of reflexivity. It helps to be aware of what frames have shaped and are still shaping reality if you are going to intervene to make the world a better place.

Reflexivity just is. In itself, it is neither a good nor bad thing. It can be either.

Framing 102 is about how reflexivity can be used for the good, at least for the good of most people, most living things, and for the beauty and bounty of the physical world that supports all life.

In all too many cases, new frames—new forms of understanding—are required to comprehend the world so as to take advantage of reflexivity and make it a better place. This is especially true when the issues confronting us, and needing framing, are complex and systemic—like global warming, the wealth gap, and many other issues that have risen to great importance over the last decade.

Let us proceed.

Systemic Causation

Studying cognitive linguistics has its uses.

Every language in the world has in its grammar a way to express direct causation. No language in the world has in its grammar a way to express systemic causation.

What's the difference between direct and systemic causation?

From infanthood on we experience simple, direct causation. We see direct causation all around us: if we push a toy, it topples over; if our mother turns a knob on the oven, flames emerge. Picking up a glass of water and taking a drink is direct causation. Slicing bread is direct causation. Punching someone in the nose is direct causation. Throwing a rock through a window is direct causation. Stealing your wallet is direct causation.

Any application of force to something or someone that produces an immediate change to that thing or person is direct causation. When causation is direct, the word *cause* is unproblematic. We learn direct causation automatically as children because that's what we experience on a daily basis. Direct causation, and the control over our immediate environment that understanding it allows, is crucial in the life of every child. That's why it shows up in the grammar of every language.

The same is not true of systemic causation. Systemic causation cannot be experienced directly. It has to be learned, its cases have to be studied, and repeated communication is necessary before it can be widely understood.

That's right. No language in the world has a way in its grammar to express systemic causation. You drill a lot more oil, burn a lot more gas, put a lot more CO_2 in the air, the earth's atmosphere heats up, more moisture evaporates from the oceans yielding bigger storms in certain places and more droughts and fires in other places, and yes, more cold and snow in still other places: systemic causation. The world ecology is a system—like the world economy and the human brain.

As a result, we lack a concept that we desperately need. We need it to understand and communicate, for instance, about the greatest moral issue of our time—global warming. The ecology is a system operating via systemic causation. Without an everyday concept of systemic causation, global warming cannot be properly comprehended. In other words, without the systemic causation frame, the oft-repeated facts about global warming cannot make sense. With only the direct causation frame, the systemic causation facts of global warming are ignored. The old frame stays, and the facts that don't fit it cannot be comprehended.

The Structure of Systemic Causation

Systemic causation has a structure—four possible elements that can exist alone or in combination. Driving a complex, systemic problem, there can be one, two, three, or all four of these elements in play. Here is how they might be explained in conversations about global warming.

A network of direct causes. (1) Global warming heats the Pacific Ocean. That means that the water molecules in the ocean get more active, move with more energy, evaporate more, and move in the air with more energy. (2) Winds in the high atmosphere over the ocean blow from southwest to northeast, blowing the larger amount of high-energy moisture over the pole. (3) In winter, the moisture turns to snow and comes down over the East Coast as a huge blizzard. Thus, global warming can systemically cause major blizzards.

Feedback loops. (1) The arctic ice pack reflects light and heat. (2) As the earth's atmosphere heats up, the arctic ice pack melts and gets smaller. (3) The smaller amount of arctic ice reflects less light and heat, and more heat stays in the atmosphere. (4) The atmosphere gets warmer. (5) The feedback loop: Even more arctic ice melts, even less heat is reflected, even more heat stays, even more ice melts, and on and on.

Multiple causes. Because of the interaction between the polar vortex and the jet stream, parts of the vortex move south into

central North America causing abnormal freezing tempera-
tures as far south as Oklahoma and Georgia.

Probabilistic causation. Many weather phenomena are probabilistic. What is caused is a probability distribution. Although
you can't predict whether a flipped coin will come down heads
or tails, you can predict that over the course of a large number
of flips, almost exactly 50 percent will come down heads and
another 50 percent tails.

Yes, global warming systemically caused freezes in the American
south. Yes, global warming systemically caused Hurricane Sandy—
and the Midwest droughts and the fires in Colorado and Texas, as
well as other extreme weather disasters around the world. Let's say it
out loud: It was causation, systemic causation! Network causes, feedback loops, multiple causes—all acting probabilistically as part of
the global weather system—have been systemically causing weather
disasters. Yes, systemically causing untold human harm and billions,
if not trillions, of dollars in damage.

Systemic causation is familiar. Smoking is a systemic cause of lung
cancer. HIV is a systemic cause of AIDS. Working in coal mines is a
systemic cause of black lung disease. Driving while drunk is a systemic
cause of auto accidents. Sex without contraception is a systemic cause
of unwanted pregnancies, which are a systemic cause of abortions.

Systemic causation, because it is less obvious than direct causation,
is more important to understand. A systemic cause may be one of a
number of multiple causes. It may require some special conditions. It
may be indirect, working through a network of more direct causes. It
may be probabilistic, occurring with a significantly high probability. It may require a feedback mechanism. In general, causation
in ecosystems, biological systems, economic systems, and social
systems tends not to be direct, but is no less causal. And because it is
not direct causation, it requires all the greater attention if it is to be
understood and its negative effects controlled.

Above all, it requires a name: *systemic causation*.

The precise details of Hurricane Sandy could not have been predicted
in advance, any more than when, or whether, a smoker develops lung

cancer, or sex without contraception yields an unwanted pregnancy, or a drunk driver has an accident. But systemic causation is nonetheless causal.

Semantics matters. Because the word *cause* is commonly taken to mean direct cause, climate scientists, trying to be precise, have too often shied away from attributing causation of a particular hurricane, drought, or fire to global warming. Lacking a concept—a frame—and language for systemic causation, climate scientists have made the dreadful communicative mistake of retreating to weasel words. Consider this quote from "Perception of Climate Change," by James Hansen, Makiko Sato, and Reto Ruedy, published in the *Proceedings of the National Academy of Sciences*:

> . . . we can state, with a high degree of confidence, that extreme anomalies such as those in Texas and Oklahoma in 2011 and Moscow in 2010 were a consequence of global warming because their likelihood in the absence of global warming was exceedingly small.

The crucial words here are *high degree of confidence, anomalies, consequence, likelihood, absence,* and *exceedingly small.* Scientific weasel words! The power of the bald truth, namely causation, is lost.

This is no small matter: The fate of the earth is at stake. The science is excellent. The scientists' ability to communicate is lacking. Without the words, the idea cannot even be expressed. And without an understanding of systemic causation, we cannot understand what is hitting us.

Global warming is real, and it is here. It is causing—yes, causing—death, destruction, and vast economic loss. And the causal effects are getting greater with time. We cannot merely adapt to it. The costs are incalculable. What we are facing is huge. Each day, the amount of extra energy accumulating via the heating of the earth is the equivalent of 400,000 Hiroshima atomic bombs. Each day!

What Journalists Can Do

Because systemic causation has mostly gone unframed and unnamed, journalists have previously been at a loss and have been driven to resort to inadequate and misleading metaphors. Charles Petit, writing in the *Knight Science Journalism Tracker* of January 7, 2014, gives a list of such metaphors. Here are some beauts:

> A weaker polar vortex moving around the Arctic like a slowing spinning top, eventually falling over and blowing open the door to the Arctic freezer . . .

> This big slug of deadly cryosphere air slipped its North Pole moorings, marauded across Canada, and swept through the eastern US . . .

> When the winds weaken, the vortex can begin to wobble like a drunk on his fourth martini . . . in this case, nearly the entire polar vortex has tumbled southward . . .

Responsible journalists can do better.

Responsible journalists need to discuss systemic causation. Certainly when discussing global warming and its climate effects, and also when discussing other systemic effects—such as those of fracking, the privatization of education, the decline of unions, and so on.

Responsible journalists also need to discuss a devastating systemic effect on our economics, recently discovered but not brought into public discourse by the press: the systemic effect of the relationship between productive wealth and reinvestment wealth.

The version of systemic causation just discussed is designed to fit global warming phenomena. In addition, there are other forms of systemic causation that we will be discussing, for example, in the study of economics. But for our purposes in this book, the most important form of systemic causation concerns the brain itself. The phenomenon of reflexivity is a form of systemic causation. And the relationship between our politics and the concept of personhood is one of the hardest cases of systemic causation to get across to the public, especially to political pundits, policy makers, strategists, pollsters, and other political professionals.

5

Politics and Personhood

Each of us has a sense of personal identity: your sense of who you are as a person. Central to that personal identity is a moral sense, a sense of what is right and wrong, what justifies our actions. That moral sense, like all that we believe and understand, is physical, built into the neural circuitry of our brains. If that changes, if the circuitry characterizing our moral sense changes, it changes our personhood. That is, it changes the kind of people we are: what we think is right and how we act.

We have seen that all politics is moral, since political policies are assumed to be right, not wrong or irrelevant. Our political divisions come down to moral divisions, characterized in our brains by very different brain circuitry. We've seen that the major moral divisions in our politics derive from two opposed models of the family: a progressive (nurturant parent) morality and a conservative (strict father) morality. That is no accident, since your family life has a profound effect on how you understand yourself as a person.

The effect of family life is complex, and peers have an effect as well. One result of that is biconceptualism. Biconceptuals have both kinds of moral circuitry in their brain, mutually inhibiting each other and applying to different issues, person by person. There is no "middle," no morally based political ideology common to all moderates.

Regardless of whether you are progressive, conservative, or biconceptual, though, your morality—your sense of what a person should be and do—is deeply connected to the way your brain triggers emotions and determines whether you feel good or bad in certain situations and about certain ideas. It is worth understanding why.

The Science behind Empathy and Morality

One of the great discoveries of neuroscience is the mirror neuron system. Simply put, that system operates in our brains and gives us the capacity to connect with others, to know and even feel what they feel, and to connect with the natural world. It is the heart of our capacity for empathy. From emotion research, we know that certain emotions correlate with certain actions in our own bodies—in facial muscles, in posture, and so on. When we feel happy, for instance, our facial muscles are prompted to produce a smile, as opposed to a frown or a baring of the teeth. We also know that the physical cues that broadcast emotion in others will usually trigger in an observer the same brain response that would accompany those physical cues of the same emotion in ourselves. That is why we can usually tell if someone else is happy or sad, or angry or bored—and why a smile is often unconsciously greeted with a smile or a yawn with a yawn.

All this is thanks to the mirror neuron system, which has circuitry connecting the brain's action centers and perception centers. As a consequence, what you see others doing is neurally paired with brain activity that could control your own actions. Muscles are activated by firing neurons, and many of the same neurons are firing whether you are performing an action or whether you are seeing someone else performing the same action. This "mirroring" allows you to see the musculature tied to the emotions of others and sense in your brain what the same musculature would be like in your body, and hence the same emotions, in yourself. In short, it allows you to feel the emotions of others! That is what empathy is about.

But this effect has further repercussions in the brain. Neuroscientists have discovered a brain overlap, too, between imagining and doing. Many of the same neural regions are activated when we form mental images as when we actually see. The same holds true for whether we imagine moving or are actually moving. That means that we have the capacity to empathize not only with someone present, but also with someone we can imagine, remember, read about, dream about, and so on. That is why we can be deeply moved by a novel or a movie, or even a newspaper story.

Neuroscientists have also shown that, when someone is in love and they see their loved one in pain, the pain center in their own brain is activated. Emotional pain is real.

Sounds simple, but there are some twists to the story, some neural complications that affect how we ultimately respond to what we see, hear, and imagine. The prefrontal cortex has regions particularly active during the exercise of judgment. These regions contain neurons that are active when we are performing some particular action and less active when we see someone else performing the same action. It is hypothesized that this gives us the capacity to modulate our empathy—to lessen it or turn it off in certain cases. The mirror neuron system thus connects us emotionally to others, but can in certain cases also distance us emotionally from others.

The prefrontal cortex is active in another neural system, too—one that I'll call the well-being/ill-being system. This is the system that releases certain hormones in your brain when you have experiences that make you feel good, and releases others when experiences make you feel bad. In essence, this system regulates whether you have a sense of well-being or ill-being at any given time. It is also the system that presumably is involved in making judgments on the basis of your imagination of what will or won't bring you well-being.

The well-being system and the empathy system can interact in complex ways. Some people feel satisfaction both when they are personally satisfied and when those they empathize with feel a sense of well-being.

Other people do not have the two systems connected in that way. (1) They may have the well-being system overriding the empathy system—with their interests overriding the cares and interests of others. Or (2) they can have a complex interaction in which they maintain their own well-being and balance it with contributing to the well-being of others. Or (3) they may be self-sacrificing, always placing the well-being of others ahead of their own well-being. Or (4) they may be part of an in-group, and may place their well-being and that of in-group members first, without empathizing at all with out-group members. This can vary, depending on what counts as a given person's in-group.

Since morality is about well-being, your own and that of others, these four alternatives define different moral attitudes.

Can the mirror neuron system be affected by inborn factors? Apparently, yes. With certain forms of autism, empathy is lessened or largely absent. In psychopaths, empathy is controlled: Psychopaths can sense what someone else is feeling, not be affected themselves, and then manipulate the other for their own benefit or enjoyment.

Can the mirror system be affected by how one is raised, by one's family life and peer relations? Does one's political morality correlate with one's capacity for empathy—that is, with the operation of the mirror neuron and well-being systems? That is being investigated, and preliminary results suggest that there is a difference between extreme progressives and extreme conservatives, with extreme conservatives showing less activation in their empathy system.

Since all thoughts and feelings are physical, a matter of brain circuitry, it is not surprising that moral sensibilities should be constituted by physical brain structures like those we have just been discussing. These brain structures form the neural basis not only of your own moral sensibilities, but also of your views on what an ideal person ought to be.

The Ideal Person

What should an ideal person be like? Conservatives and progressives have largely opposite views, given their different views of morality. Biconceptuals have different views as well, depending on how their moral views are divided up: biconceptuals who are largely conservative will tend to have a conservative view of what people should be like, and biconceptuals who are largely progressive will tend to have a progressive view of what people should be like. Or, biconceptuals that are less extreme may believe that an ideal person is biconceptual in the same way they are, with the same distribution of conservative and progressive views.

The progressive (nurturant parent) moral system maintains a delicate balance between the empathy and the personal well-being systems. At its core is empathy for others and the responsibility to act on that empathy, but it is modulated by the proviso that you can't take care of anyone else if you're not taking care of yourself. That is, it centers on empathy and includes both personal and social responsibility.

The conservative moral system centers on the well-being system— on personal responsibility alone, on serving your own interests without depending on the empathy of others to take care of you and without having empathy and responsibility for others.

There are nuances, but this gets at the heart of the difference.

Empathy versus Sympathy

Empathy and sympathy both involve the capacity to know what others are feeling. But unlike empathy, sympathy involves distancing, overriding personal emotional feeling. Someone who is sympathetic may well act to relieve the pain of others but not feel the pain themselves. The word "compassion" can be used for either empathy or sympathy, depending on who is using the word. For example, George W. Bush, in first announcing his run for the presidency, called himself a "compassionate" conservative, citing the book by Marvin Olasky, *The Tragedy of American Compassion.*

Olasky and Bush's take on compassion and conservatism point to a central difference between progressives and conservatives. Progressives tend to believe that society as a whole has a responsibility to aid those in real material need and that the government should be a major instrument, with support from taxes. Conservatives tend to prefer charity, delivered through nongovernmental organizations, and tend to believe that real help for most people in material need is a refusal of aid, to give them an incentive to help themselves. Hence the conservative motto: It is better to teach someone to fish than to give him a fish to eat. Incidentally, charity for the "deserving" few costs a lot less than taxes to provide resources for the benefit of all.

This dichotomy leads to two very different ideas of what an ideal person should be like, and how our politics should be arranged to produce a version of the ideal person with the "right" moral system, whether purely conservative, purely progressive, or the right combination of the two.

Reflexivity and Personhood

At this point we have to ask The Reflexivity Question for Personhood: Can linguistic framing change the kind of person someone is? The answer seems to be yes, though possibly not in extreme cases. And of course it may depend on age and circumstances. But such changes do appear to have happened over the years—so far as I can tell, mostly with biconceptuals. Extreme conservatives (estimated at about 25 to 30 percent of the US population), it appears, cannot be changed by reframing and setting up an effective communication system that operates full time, not just at elections. Yes, this means that some people cannot be "reached" (an inaccurate progressive metaphor) or "woken up" (another inaccurate progressive metaphor).

Consider a moderate progressive who is partly conservative. She hears conservative language and conservative arguments over and over, day after day for years—in the media or with friends or both. The conservative language will activate the conservative moral system, making it a bit stronger every time the language is heard. As the conservative circuitry in her brain becomes stronger (the synapses strengthen), the more likely it is that her views on issues will change from progressive to conservative. The result may be a shift within the brain from a person who is partly conservative to a person who is mostly conservative. I believe that this has actually happened in many cases.

That is the power of the conservative messaging system: It is reflexivity in action. Over time, someone's very personhood can change, and with it her ideal of what other people should be. And, of course, who they should vote for.

The other conservative use of reflexivity depends upon getting those votes. Once in office, conservatives can not only say that government cannot work and has to be minimized and privatized, but by being in the government, they can also stop it from working, thus creating a self-fulfilling prophecy. How? By cutting taxes, by cutting funding, by passing laws, and, in the Supreme Court, by reinterpreting laws.

In contemporary America, politics and personhood are inseparable—and apparently moving in a conservative direction. To change that direction, progressives need to understand the role of the brain and of communication systems in the process.

Politics and Personhood at the Founding

When the United States was founded, politics and personhood had come together, but in the progressive direction.

Historian Lynn Hunt at UCLA goes through the history in detail in her book *Inventing Human Rights: A History*. She starts with the defining passage of the Declaration of Independence:

> We hold these truths to be self-evident, that all men are created equal, that they are endowed by their Creator with certain unalienable Rights, that among these are Life, Liberty, and the pursuit of Happiness.

If these rights are self-evident, she asks, why does Jefferson have to say that they are self-evident? And when did they become self-evident?

Hunt, a former president of the American Historical Society, studied the writing and culture of France, England, and the Thirteen Colonies. She shows that those ideas were not there in the 1600s, and came into existence in the mid-1700s, mainly after 1760, when Western Europe and the States were swept up in a major cultural change. That change can be seen in the period's novels, like Jean-Jacques Rousseau's *Julie*, the biggest bestseller of the century, with seventy editions between 1761 and 1800. *Julie* was written as a collection of intimate letters between two lovers. Readers identified deeply with the emotional lives of the characters, whose psychological states were revealed and developed from letter to letter, arousing empathy for the plights of ordinary people. Between 1760 and the 1780s, such novels multiplied, laws were passed ending torture by the state as being inhuman, portraits showing the individual characteristics of their subjects started to be painted, manners changed to increase personal control over one's body (e.g., blowing your nose into a handkerchief), and the idea of individual autonomy came into existence in a rush.

These changes were propelled by empathy, by identification with the problems and plights of ordinary people, feeling what the characters felt, seeing such plight around them, and propelling legal and governmental change. By 1776, human rights became "self-evident" via the development of empathy for one's fellow citizens. Such empathy formed the basis for a union of states, and American democracy.

Historian Danielle Allen, of the Institute for Advanced Study at Princeton, has taken the study of the Declaration of Independence one important step further in *Our Declaration: A Reading of the Declaration of Independence in Defense of Equality*. It's a thorough reading of the declaration, though the central passage again is the classic one on self-evident truths. But Allen, going through the original copies, argues that the period at the end of that passage was not there in the original document: It was inserted later. Her case is backed up by the grammar of what follows. Here is the passage as a whole, with the original punctuation:

> We hold these truths to be self-evident, that all men are created equal; that they are endowed by their Creator with certain unalienable rights; that among these are Life, Liberty, and the pursuit of Happiness; that, to secure these rights, governments are instituted among Men, deriving their just powers from the consent of the governed; that whenever any form of government becomes destructive of these ends, it is the right of the people to alter or to abolish it, and to institute new government, laying its foundation on such principles, and organizing its powers in such form, as to them shall seem most likely to effect their safety and happiness.

Allen argues that, with the period, the self-evident truths end with life, liberty, and the pursuit of happiness, what I've described above as coming from the well-being system, but lacking the idea of citizens establishing and working through a government on the basis of empathy for the well-being of all. The passage ending with life, liberty, and the pursuit of happiness is about freedom, but what follows is about equality, with the central role of the government to secure it.

The grammar—the succession of *that* clauses and the plural ("to effect their safety and happiness")—shows that the passage goes beyond the pursuit of happiness to the role of government in guaranteeing equality in the unalienable rights.

Allen is right that this is a big deal. Her point is not just about a period. It is about our political division. As I describe in my book *Whose Freedom?*, progressives and conservatives have very different views of freedom. Conservatives talk about their version of freedom, which does not include either equality or the role of government in securing it. They also attribute their view to the Founding Fathers. And though conservatives seek radical change, they use the term "conservative" as if it conserved the values of the founding of our nation.

At issue is what freedom is supposed to mean, what democracy is supposed to mean, and what personhood is supposed to be.

6

The Private Depends on the Public

From the beginning, America provided public education, public hospitals, public roads and bridges, an army to protect the union, a legislature to make laws regulating and maintaining the union, an executive to carry out those laws, a justice system to enforce those laws, a national bank, a patent office, means to promote interstate commerce, and above all, a system for the public to choose those who govern us. Without such public resources there could have been no satisfactory private life and no functioning business community in America—and no democracy!

This tells us something deep and crucial about American democracy, and Western democracy in general. American democracy has grown out of the idea of a union—a coming together of citizens who care about each other and therefore care about their nation as a whole. America has worked as a democracy because enough Americans have taken responsibility for each other—that is, for the nation—using their government to provide public resources for all, enough and of the right kind to provide decent private lives for most of our citizens.

Understanding this requires noticing and appreciating those public resources, appreciating the civil servants who provide them, and understanding that we as citizens take on the responsibility, both through paying for them and providing political support.

It is even more true today that the private depends on the public. We, as citizens working though our government, have provided much more—an electric grid, public universities, an interstate highway system, publicly funded scientific research that has resulted in the field of computer science and all computer technology, satellite communications that make telecommunications and the Internet possible, modern medicine, airports and an air traffic control system, pilot training through the air force, a center for disease control and a food and drug administration, an environmental protection agency, a national park and national monument system, a public resource management system, a civil service to replace the corrupt old spoils system, and on and on. And perhaps the greatest public resource of all is a public system for managing and guaranteeing the functioning of all of these public resources: a government—a system of governing—up to these tasks.

Without all of this, the blessings of modern American private life and private enterprise would not be here. The private depends on the public. Public resources make private life possible.

It doesn't take much to see this. The evidence is all around us every day. There used to be signs on public projects: "Your tax money at work!" But such signs appear no more and the most basic truth of our democracy goes largely unspoken. Why?

Progressives take it for granted, as part of their moral and practical assumptions, like breathing or noting that the sky is blue. This is an important fact about how brains work. Some ideas and some knowledge are so deep that they rarely if ever even come to consciousness. Nobody goes around saying things like, "People breathe," or "You have a nose."

But for conservatives, the very idea that the private depends on the public is anathema—immoral. Conservatives have a different view of responsibility. Whereas progressives believe centrally in empathy (caring about their fellow citizens), both personal and social responsibility, and a commitment toward doing their best toward those ends, conservatives believe only in personal responsibility.

This yields a completely different view of democracy, that democracy provides what they call "liberty"—the ability to seek one's own interests without the responsibility of others to help them, without any responsibility to help their fellow citizens, and without interference from the government.

This is a moral conviction, as deep in the conservative brain as the progressive moral vision is in the progressive brain.

Again, I say "brain"—not "psyche" or "mind"—for a deep and vital reason: All thought is physical, carried out by the neural circuitry in one's brain. Thoughts don't just float in air. As a result, you can only understand what your existing brain circuitry allows you to understand. The fundamental frames through which you understand the world are physical. Your moral identity is as much a physical part of you as your lungs or your nose. You can only make sense of what your brain allows. If the facts don't fit what your brain physically allows, the brain circuitry stays and facts are either ignored, dismissed, ridiculed, or seen as a form of immorality to be fought against. It is a fact that the private depends on the public—perhaps the most central fact of American democracy—and yet strict conservatives either can't see it or see it as a form of immorality so fundamental that it must be defeated at all costs.

This is a major part of what is driving the divisiveness of our country and the conservative move to make our government dysfunctional. It is behind the conservative moves to privatize as much of government as possible: to privatize education, public health, public safety, water resources, regulation of business practices, much of national defense, and on and on.

Can brains change? Can the brains of enough Americans change so that this basic fact of our democracy can be comprehended and appreciated?

In many cases, no. But in the overwhelming majority of cases, yes. What makes it possible is biconceptualism.

The Brain and Constant Public Discourse

Biconceptualism is a fact about the brain. A great many people have in their brains versions of both progressive and conservative moral values in all sorts of combinations. And they apply these different moral values to different issues.

Remember that there is no ideology of the moderate—no set of views held by all moderates. A moderate progressive has mostly progressive views, but also some conservative ones. A moderate conservative has mostly conservative views, but also some progressive ones. But there is no single set of policies that defines a "middle."

Progressive and conservative worldviews contradict each other. Both are characterized in the brain via neural circuitry. How can you have contradictory neural circuits in the same brain? Easy. The answer is mutual inhibition, a very common kind of brain circuitry: When one circuit is active, it turns off the other. Which is turned on at a given time is a matter of context. Someone with both worldviews applies them to different issues in different contexts, resulting in the brain circuits for the different values unconsciously and automatically switching back and forth depending on the issue. That's what it means to be a biconceptual.

It is often these biconceptuals who, as voters, are the target of intense campaign attention. But Republicans understand how to appeal to them better than Democrats do. Recall that all politics is moral, and that a voter's implicit sense of morality is absolutely central to a voter's identity. Since biconceptual voters have both moral systems—mostly one but partly the other—conservatives need to keep their voters and attract the partly conservative moderate Democrats. Conversely, progressives need to keep their voters and attract partly progressive moderate Republicans. There is an honest strategy for achieving this goal, and also an Orwellian one. The honest strategy is to use only your language and avoid using the other side's language. That will maximally activate your moral system in the moderates on the other side. The Orwellian strategy involves using the other side's language in an attempt to "reach" those with moderate or opposing views. But if you use Orwellian language, you will be activating the other side's moral system, making it stronger, and shooting yourself in the foot.

However, some political organizations use Orwellian language. For instance, an hour before writing this, I received a robocall from the "Center for Worker Freedom" asking me to support a measure that they said would support worker freedom. They are an anti-union organization and the measure is anti-union, though the telephone message didn't mention that. They were trying to fool Democrats into supporting an anti-union measure out of ignorance.

It is vital that progressives understand why just citing the facts doesn't work, and why attention to public discourse must be constant, not just focused on elections. Here are the basics that progressives need to understand.

There is a crucial logic to the way the brain works with respect to public discourse. Here are ten key points to that logic.

1. The more a brain circuit is activated, the stronger its synapses get.
2. The stronger its synapses get, the more likely it is to fire and the stronger the firing is.
3. When two circuits inhibit each other, the stronger one circuit gets, the weaker the other gets.
4. Suppose two mutually inhibitory circuits apply to different issues. As one gets stronger and the other gets weaker, the more likely it is the stronger one will start applying to more issues and the weaker one to fewer issues.
5. Language changes the strength of those circuits. Conservative language activates circuitry for the conservative worldview; progressive language activates circuitry for the progressive worldview.
6. Imagery fitting one worldview or the other matters as much or more.
7. Frequency of language use and imagery matters. The more frequent the language use or imagery, the more strengthening occurs.
8. Journalists are trained to use the most frequent language in public discourse.
9. The conservative turn in America has come from the constant use of conservative language in public discourse. So much so that progressives have often adopted conservative language, thus helping the conservative cause.
10. Because of the effect of language and imagery on the brain, the constant use of one ideology's language over the other's has an enormous effect on our politics.

Conservatives have been doing a better job at getting their language into public discourse. The enormous conservative communication system has done its job well, especially on the central issue of our democracy—that the private depends on the public.

The central conservative strategy to minimize, or even eliminate, public resources has been to eliminate the money that funds public resources—taxes! Taxes for the wealthy have been cut by conservatives, who have defended huge tax loopholes, and have even drastically cut funding for the IRS so that there are not enough IRS workers or modern computers for the IRS to monitor tax evasion—mostly by the wealthy. Since the 1970s, the concept of taxation has shifted from the source of needed, and often revered, public resources to the idea that taxation is a burden—an affliction in need of "tax relief."

The constant talk of taxation as an affliction and a burden has led biconceptuals to "switch" to viewing taxation as a burden rather than something that makes our private lives possible or that creates a base from which corporations prosper. While conservatives drive these frames home, progressives don't realize that they have to drive their own frames home—and only later realize that the conversation has completely changed. "Suddenly" not just conservatives are talking about the burden of taxes instead of the value of public services, but so are the media, so are moderates, and, eventually, the "tax relief" language works its way into even progressive discourse. The term *Tea Party* was chosen to make it sound patriotic to oppose taxation.

The only progressive who has succeeded in getting across the idea that the private depends on the public is Elizabeth Warren, who has argued it repeatedly, and did so with special success when she was running for the Senate in 2012. At one point in his presidential campaign, President Obama tried the argument, but messed it up when he tried ad-libbing in a public talk and said, "If you've got a business, you didn't build that. Somebody else made that happen." Conservatives jumped on his remark and attacked him viciously for it. He could have fixed it by making the point correctly the very next day and every day thereafter, getting the idea out into public discourse, and allowing the press to do its job and present the overwhelming evidence for it. But the president was too timid and dropped it, missing a major opportunity to change public discourse.

Can progressives turn this around? Yes, but not without serious conscious commitment. The president and every progressive candidate, office holder, and public figure of any kind can start now: say it right, over and over. Connect the private-depends-on-the-public concept to something that conservatives will understand: freedom. Public resources allow for freedom in case after case, opening up all kinds of opportunities in life. It is the freedom that public resources afford that make them central to democracy.

Saying it right—and saying it over and over—is advice that can be applied to issue after issue.

★★ PART III ★★

FRAMING FOR SPECIFIC ISSUES

7

Freedom Issues

One of the major mistakes made by the Democratic Party is to focus on election campaigns but not on the constant framing of public discourse. All politics is moral. Voters vote on what they implicitly, automatically, and unconsciously believe to be right. In short, elections have everything to do with how biconceptuals have adopted the moral vision of one side or the other. And elections depend on the language voters hear and the images they see every day—not just during the campaigns.

Democrats tend to address interests: kitchen table economics and the objective facts affecting the interests of middle class and poor voters. Yet, poor conservatives and biconceptuals regularly vote against their interests. Many Tea Party voters are poor or made poorer by conservative policies. But the conservative message machine is relentless and extends almost everywhere. Conservative messaging dominates everyday public discourse. And the domination of everyday public discourse—at least as much as the domination of electoral discourse—determines our political realities.

Conservatives have come to own the words *freedom* and *liberty*. These words weigh heavily in the conservative vocabulary. These words are among the most powerful in our politics because of the centrality of the concept of freedom to democracy. Conservatives have no right to that ownership.

Words have contestable meanings, and the word *freedom* means very different things to progressives than to conservatives. As I pointed out in *Whose Freedom?*, freedom is a contested concept. Conservatives and progressives use the word to opposite effects.

As we have seen from a careful reading of the original Declaration of Independence documents, the progressive meaning is at the heart of our democracy and it is time to take it back. Most of the issues in public discourse, both in elections and in everyday decision making, come down to issues of freedom.

Health Care

Comedian Jimmy Kimmel sent an associate out onto a Los Angeles street corner with a microphone to ask passersby a simple question: Which do you like better, Obamacare or the Affordable Care Act? The overwhelming majority said they didn't like Obamacare, but thought that the Affordable Care Act was a good idea. Most did not know that they were the same thing. After all, different names typically refer to different things.

How did they get the negative impression of "Obamacare"?

After Obama was elected in 2008, but before he was inaugurated, he had a pollster do a survey of which provisions for his health care act would be most popular. A group of those provisions came in between 60 and 80 percent in popularity. They were the familiar ones: no preconditions, no caps, your college-age child can be on your plan, and so on. These became the main provisions of the plan. The assumption was that if all the main provisions were popular, the whole plan would be popular. In other words, the popularity of the plan should depend on the popularity of the provisions of the plan.

No conservatives attacked those popular provisions. There was no conservative movement in favor of preconditions or caps, or against having your college-age child on your plan.

Instead, conservatives understood that politics is a matter of morality and decided to attack the plan on moral grounds. They chose two moral domains: Freedom and Life. On Freedom, they attacked it as a "government takeover." On Life, they said it contained "death panels." And they repeated "government takeover" and "death panels" over and over, month after month. And every time the president said "It is not a government takeover," he used the words *government takeover*, which activated the idea of a government takeover in the brains of listeners, thus reinforcing the conservative attack.

The conservatives also never used the name Affordable Care Act. Instead, they invented their own name—Obamacare, taking the emphasis off the affordability of health care and associating *Obamacare* with *government takeover* and *death panels*. The press, quoting the attacks, used the term Obamacare and not the clunky name Affordable Care Act. Obama eventually tried, in vain, to turn this to his advantage—saying it meant "Obama Cares." But it was too late. The conservatives had given the name the meaning they wanted by frequent enough repetition.

The president and members of the administration counterattacked with lists of facts: the provisions of the plan. It didn't help. The president went on TV with a laundry list of provisions. It didn't help. His adviser, David Axelrod, sent a memo to Organizing for America's email list, roughly 13 million supporters strong, asking them to speak to their friends and neighbors in support of the president's plans. He said there were twenty-four things to remember, but just to make it "easier," he divided the list of facts into three groups of eight!

Any cognitive scientist could have told him no one was going to remember the three groups of eight, and I have never met anyone who has.

The conservatives won the framing war of 2009, and it helped strengthen the nascent Tea Party movement as Tea Partiers were deployed to go to town meetings around the country that summer and repeat *government takeover, death panels,* and *Obamacare.*

If the president had understood the conservative framing tactic, he could have undercut it in a simple way. He could have adopted the same two moral issues, Freedom and Life, from a progressive perspective.

If you have cancer and you don't have health care, you are not free. You are probably going to suffer and die (a Life issue). If you are in a car accident and suffer multiple injuries and don't have health care, you are not free—you may be disabled for life, or die. Even if you break your leg, do not have access to health care, and cannot get it set, you are not free. You may never walk or run freely again.

Ill health enslaves you. Disease enslaves you. Even cataracts that rob your vision and can easily be healed by modern medicine will enslave you to blindness without health care.

Healthy food is also a freedom issue. Much of big agriculture produces unhealthy food, especially processed food, sugary food, food with unhealthy additives, meats with hormones and antibiotics from animals raised on pesticide-treated feed, and so on. Access to healthy food is a freedom issue.

And when conservatively run states turn down funds for Medicaid, that is a freedom issue—both for the people who are being denied health care, and for everyone else to whom a curable disease can spread when health care is denied to a significant number of the people they interact with every day.

Freedom issues are powerful issues.

Education

Conservatives want to eliminate public resources as a moral issue. In their view, they are given for free and therefore take away personal responsibility and the incentive to work. Education is a main example. In conservatively run states, like Wisconsin, funding for public education has been severely cut.

The conservative movement against public school education offers the alternative of charter schools, religious schools, and private schools. Charter schools are schools paid for publicly but run privately—very often by for-profit corporations.

The CREDO study at Stanford in 2013 found that about 75 percent of charter schools have results that are worse than, or no different from, traditional public schools. A small percentage of charter schools do have better results. But since funds for charter schools are taken from public school budgets, charter schools tend to drain money from public schools and make public education worse on the whole, even for the best public schools.

Moreover, charter schools have no accountability to local school districts or the public. A consequence in Texas, for example, is that charter schools tend to debunk evolution and science and teach creationism. In Michigan, 80 percent of schools are now charter schools, and they are doing no better at educating children in poverty than public schools.

The conservative framing is that public schools are "failing" and that vouchers for religious or private schools give parents "choice." Those vouchers tend not to pay for high-quality schools, so that poor families that receive them tend not to get high-quality education for their children. But for wealthy parents, the vouchers represent public support for the wealthy and a cut in support for those who lack wealth.

The conservative attack on public education is being felt drastically in higher education. Conservatives in state legislatures are cutting funding for higher education, with two horrendous consequences. State-run colleges and universities used to be the gateways to education for poor and lower-middle-class students. As conservatives cut state university budgets, the schools, to stay in business, have had to raise their tuition, pricing higher education out of reach for a great many of these students. Students' only alternative has been to borrow money, which raises the second problem: student debt. At a time when banks can borrow money at 1 percent interest, students have to pay 8 percent interest on their loans, which burdens them with many years of loan payments after they graduate. That makes it harder for them to afford getting a post-graduate education or starting a family. Present calculations are that if the government forgave all student loans, it would boost the nation's economy far more than the cost of the loans. Nonetheless, conservatives are against both loan forgiveness and dropping the interest on student loans to the same rate that banks pay.

Whether at the level of pre-school, K–12, or higher education, the conservative move is to reduce or end public education—as part of the move to end public resources in general.

Education is a freedom issue. But that is not now being said in public discourse. Without education you are not free in many, many ways. Education tells you about the world and the possibilities in life. If you don't know what is possible, you cannot even set goals. Education isn't just about filling your head with facts; it's about teaching you to think, to notice, to be critical, to act rationally, to be practical, and to get access to facts for yourself. Education gives you skills, the ability to do things you couldn't do otherwise. Yes, educated people have more economic potential—and money can make you free in many ways—but the freedom education offers goes well beyond money. It opens the possibilities for connections to the natural world, for an aesthetic life, for a life of ideas, for an understanding of what is going on around you, and for an understanding of yourself. And it gives you the knowledge and the opportunity to be a productive citizen, to contribute to your own freedom and the freedom of others via political and social engagement.

If education in general is a freedom issue, public education, on the whole, is an even more powerful freedom issue, with two important components:

- **Public education is publicly accessible.** It gives educational access to more people, and so increases their freedoms. It also allows individuals to understand the full range of people, and therefore opens up human relationships and the possibility of empathizing with and understanding more people.
- **Public education is publicly accountable education.** It prevents the narrowing of what is taught when private interests determine what is taught.

Moreover, systemic causation further makes education a freedom issue, because many of our major educational issues are due to poverty.

Poverty all too often means that parents who have to hold down multiple jobs cannot properly raise their children—not reading to them, not raising them to respect education, not being able to get them out of unhealthy environments. Poverty means children going to school hungry in the morning and not able to concentrate on classes. And poverty and lack of education replicate themselves. In a large range of cases, the failure of students to learn has mainly to do with a national economic failure and not with inadequate teachers or schools.

Poverty

Poverty is a freedom issue. It is obvious. People who are poor have a lot less freedom than people who are rich.

As we have seen, people who are poor have less access to health care and education than people who are rich.

But there are many more freedoms eroded, or altogether lost, due to poverty. Housing is better for the rich than the poor. There is no homelessness among the rich. Neighborhoods are better for the rich than the poor. The ability to relocate or travel is far easier for the rich than the poor. Food is better for the rich than the poor. Social connections are better for the rich than the poor. Better jobs are available for the rich than the poor.

All of these are freedom issues: If you are homeless or cannot find a decent place to live for yourself and your family, you are oppressed and limited, you are not free. If you cannot relocate or travel when you need to, or want to, you are limited in your freedom. If you cannot eat properly, you are not free. If you are not able to connect with people, you are not free. If you are not free to find a job and work, you are not free.

In virtually every dimension of life, being in poverty without being able to escape it is a freedom issue.

Yet many people in poverty often vote for measures that make their lives worse, not better, because continual conservative framing has activated a conservative worldview even in those whose lives could be essentially ruined by it.

Conservatives see being poor as a personal failure, a failure of individual responsibility. But the reality is that poverty curtails freedom. There is a reason why people speak of being "trapped" in poverty. They are.

Again, in our democracy, the private depends on the public. Do we care about whether our fellow citizens are free, or not?

Discrimination: Race, Gender, and Sexual Orientation

By now, our history has made clear that racism is a freedom issue. It can impose poverty, lack of education, ill-health, and worse. As we saw with Trayvon Martin, it can get you killed in some states.

Our recent history has made it clear that homophobia is a freedom issue. It is as normal to be gay as it is to be left-handed. Freedom to marry for people who happen to be gay is as much a matter of love and commitment as it is for people who happen to be straight, and a denial of marriage or other rights on the basis of whether you are straight or gay is a freedom issue. That is becoming clearer all over America.

The first edition of this book played a role in the adequate framing of the deepest reality of gay rights. The prior arguments were about rights of inheritance, of hospital visitation—discrimination in monetary and social matters. These were practical issues. What this book made clear was that the issue was fundamentally moral—a matter of love and commitment. All people should be free to marry whoever they love and want to be committed to for life. Progressives began using that message more and more frequently beginning a decade ago, and we have seen the right to marry progress by leaps and bounds.

Framing the truth at the deepest moral level matters.

What have been called "women's issues" are also freedom issues, and these have not been adequately framed as such. In general:

- **Body control.** The right of human beings to control their own bodies is a freedom issue.
- **Respect.** The right of human beings to be treated institutionally with respect as a human being is a freedom issue.

Women are human beings and have a right to control their own bodies. When that is denied, they are not free. Control over a woman's body arises in a wide variety of cases:

- **Sex education.** For women especially, sex education is required for control over one's body, since women need education about menstruation, sexually transmitted diseases that can affect future childbirth, how sex can lead to pregnancy, and how reproduction can be controlled.
- **Control of reproduction.** Reproduction occurs through women's bodies and affects those bodies in a great many ways. Women need to be in control of whether or when they reproduce. Thus access to family planning advice, birth control methods, and abortion are issues of control of a woman over her body.
- **Forced ultrasounds and attacks on family planning.** Forcing a woman to undergo humiliation in order to exert control over her own body is a freedom issue. For example, forcing a woman, as in Texas, to have a mostly male-administered ultrasound

twenty-four hours before an abortion, or allowing anti-abortion advocates to hound her on her way to a clinic, is a freedom issue for women. Passing laws that make it impossible to keep family planning clinics open is also a violation of women's freedom.

- **Humiliating victims of sex crimes.** A free woman has control over her own body. Sex that violates that control includes rape, drugging a woman in order to have sex, exerting physical or psychological force to have sex, and so on. Police and courts who humiliate a woman who has been raped are violating her freedom.

These are all freedom issues. They are rooted in circumstances that apply to women, but they are special cases of the freedom of all human beings to control their own bodies.

There are also circumstances where women are not being treated like other human beings on an institutional level—in significant ways:

- **Equal pay for equal work.** This is not just an equality issue. It is an issue of whether women are being treated like any other human being would, or should, be treated.
- **Equality in the rating of ability for a position in an institution.** In a free society, gender should make no difference in whether or not a person gets a job, a promotion, admission to an academic program, nomination for political office, and so on.

These, too, are freedom issues. You are not free when you are not treated like other human beings with respect to how you function in an institution.

Equality and freedom are not separate issues. Discrimination is a denial of freedom. Freedom is more general. It has to do with a clear path (no one standing in your way or placing obstacles) or with possessions you have a right to. It is freedom that is at the heart of democracy. And it is freedom that concerns everyone who has needs, dreams, and goals.

The present Democratic framing is the War on Women. I don't know if it is a good money-raising tactic. But it is not effective framing beyond strong feminist progressives. Strict father morality is partly about preserving male authority over women by claiming protectiveness and support of women—anything but a war against them. Conservative women, too, tend to see male authority as protectiveness, or support for motherhood as the basic female function.

The War on Women works for feminist progressive women, who correctly see their values as under attack. But it doesn't work so well for conservative or biconceptual women. Freedom, on the other hand, allows women to decide for themselves, whatever their views on abortion, contraception, and sex education.

Unions and Pensions

Workers are profit creators.

Conservatives like to speak of wealthy company owners and investors as "job creators," that they "give" people jobs, as if they just create jobs as gifts for people who are out of work. That is nonsense. The truth is that workers are profit creators, and that no one gets hired unless they contribute to the profit of owners and investors.

It is basic truth. Workers are profit creators. But who says it? How many times, if any, have you heard that truth? It is an important truth; it reframes the issue of jobs from the perspective of the contributions of those who work.

As we discussed earlier, a pension is a delayed payment for work already done.

This is the most fundamental truth about pensions, and it is almost never said. It is an unframed truth.

When you take a job and a pension comes with it, that pension is part of your pay, part of your conditions of employment. It is common for workers to forego higher current pay if there is a significant pension, since the pension is money to live on when you can no longer work. It is part of the employment contract.

The idea behind pensions is that a company can pay less in salary, take the remaining money and invest it, assuming it can invest it at a higher return than the worker could, then make a profit on the investment return when the pension is later paid.

An additional idea behind pensions is that they keep employees loyal to the employer, and so employers save money on having to train new workers, and in addition they can retain workers who know the business and so can be more efficient than new workers.

In short, a pension is anything but a gift to an employee. It is earned. And it is set up to profit the employer as well as the employee.

Unfortunately, money for pensions is often misappropriated or mishandled by the institutions. It may be badly invested or used for some other purpose, like paying dividends to stockholders or salaries to management. So when a company (say, General Motors) or a city or state says to its employees that it cannot "afford" to pay pensions, they are engaging in theft and the thieves should be prosecuted.

The money has been earned. If it has been used for some other purpose, it has been stolen. If it has been badly invested, then the investment loss is the company's, and the pensioners should have a claim on the company's assets.

Unfortunately, framing enters in here. Pensions and health care are called "benefits," as if they are generous gifts to employees. They are not gifts. They are earned as deferred payments for work done. When a company tells its employees that they can no longer afford such "generous benefits" and will have to cut them, it is a framing lie. Either there has been theft or misinvestment or mismanagement. "Benefits" are earnings, period.

Pensions and benefits are freedom issues. In a free society, there is a justice system that punishes thefts, adjudicates contracts, and in the case of misappropriation of funds, permits lawsuits to make a claim on assets to make up for losses and the costs of the lawsuit—emotional and monetary. To the extent that there is no such justice system, people with pensions and benefits that have been taken from them are not free.

Large companies—and some small ones—have two kinds of employees: the assets and the resources.

The "assets" include major management and especially creative or skilled people whose special creativity and skills are necessary to the company's success. They are part of the stock value of the company. They are hired by "headhunters" and command high salaries and golden parachutes—high pensions and compensation packages.

The "resources" are interchangeable workers that can be hired from an employment pool. They are hired and managed by the Human Resources Department. Just as resources like gas or oil or steel are purchased as cheaply as possible, so, too, are human resources purchased as cheaply as possible. Since pay scales often match skill level, they tend to be hired at the lowest possible skill level and at the lowest cost. When unemployment is high and the employment pool is large, companies can offer less in salary and "benefits" and still get appropriate human resources, while maximizing profits and payments to "assets."

Unionization is a freedom issue.

Companies that are hiring human resources, in general, have much more power than individuals seeking such a job. When the company is large and there is a big human resource pool, workers seeking jobs have to take what is offered—or the job will go to the next person in the pool. This includes not just salary and benefits, but also working conditions—job safety, working hours, overtime, and so on. The employee is serving on the company's terms, and often at the company's whims.

In capitalist economic theory, employment is a transaction in which the employer buys the labor of the employees and the employees sell their labor to the employer. Hence the term *labor market*. It is assumed in economic transactions that both will seek the best deal. Unions create the best deal for the resource-employees.

Unions function to equalize the power of the company over the employee. Short of outsourcing, companies cannot function without any resource-workers at all. If the company is unionized, then all the workers as a group have bargaining power that a solitary worker does not have.

The alternative—taking whatever the company offers to the individual—might well be called corporate servitude or wage slavery. As the power of unions has declined, the wages of resource-workers have not gone up in thirty years. Over the same time, the wealth of wealthy investors and corporations has skyrocketed without more being produced.

The decline of unions has meant a decline for most citizens in their share of their nation's wealth, and with it a decline in all the freedoms that wealth brings.

Unionization is a freedom issue, and needs to be understood as such. But the failure to say it out loud and repeat it as often as possible allows conservatives to form organizations like the Center for Worker Freedom, as if unions were taking away freedom, and to speak of "Right to Work" laws, as if unions were taking rights away instead of granting you freedom from corporate servitude and wage slavery.

Immigration

America is a country of immigrants. Many of them have been refugees, either refugees fleeing from brutal oppression or economic refugees fleeing from equally brutalizing poverty. They have come here for freedom.

My own grandparents were such refugees—and if you are not Native American, your ancestors most likely were too. Upon arriving in America, my grandparents became Americans in the best sense of the word: hard-working, raising their families, highly ethical, and loving and appreciating this country. I suspect that your ancestors were like that as well.

The issue of "immigration" is about a new generation of such refugees. President Obama, in a speech on June 22, 2012, at the National Association of Latino Elected and Appointed Officials conference in Florida clearly and beautifully stated his moral understanding of the issue. His words showed that the current wave of refugees, referred to as "undocumented immigrants," are in many ways already citizens—they contribute enormously to American society and the American economy through hard work, they love the country they live in, they are patriots, they share their lives with other Americans every day, they take on individual and social responsibility. The president offered more than just freedom; he offered appreciation. They have earned not just recognition as Americans, but our gratitude as well for all that they have contributed through hard work, often at low pay.

They are fine Americans already and, through the lives they have been living as Americans, have earned the documentation that other Americans have gotten just by being born, without earning it. This is a moral narrative that tells a truth and needs to be repeated. But it rarely is.

There are two metaphors, one liberal and one conservative, that do not do the refugees justice. The liberal metaphor is the Path to Citizenship, as if citizenship should be the end of a long, hard journey, with little granted along the way, with long years in limbo, and legal residency only to those who act as ideal citizens and either go to college or serve in the military. The DREAM Act, which would allow such access to the American Dream, doesn't have the right name. It makes these de facto citizens into those who can only dream, as if they are not acting every day just as citizens—the "best" of our citizens—act. At the very least they are earning, and deserve, a minimum along the way: health care, decent housing, decent working conditions, a living wage, and access to education for themselves and their children—and the right to a driver's license. They deserve not just freedom, but gratitude.

The conservative metaphor shows anything but gratitude. It is the Criminal metaphor. In fleeing to America, often risking their lives to come, these refugees trespassed; they did not have documents, which is not within the law. Conservatives have therefore branded them as "criminals"—"illegals"—as if they are committing crimes every day when they are actually mowing lawns, cleaning houses, taking care of children, picking vegetables and fruits, cooking your meals, working on construction sites, and, whenever possible, using all the skills they have to their advantage and to ours. Their children are studying in schools and helping out at home.

But because they often have brown skins, are impoverished, and speak Spanish, they are discriminated against. Conservatives want to jail them and deport them. Being brown-skinned, Spanish-speaking, and poor, and not being born American, they fall low on the conservative moral hierarchy: They are seen as less moral. They are discriminated against for their color and language and blamed for their poverty.

The issue for Americans is empathy: Do we care for those fellow human beings who are functioning as our fellow citizens? Or do we treat them as lesser beings, not worthy of the freedom they are earning day by day? The issue for those who have come here to escape the brutality of oppression and poverty is freedom.

This is especially true for the tens of thousands of children who have crossed the border, sent by parents or fleeing on their own from human traffickers, gangs, and death squads in Guatemala, Honduras, and parts of Mexico who are murdering, harming, or kidnapping children. Under an executive order signed by George W. Bush before he left office, these children have to be taken care of reasonably well by the US government, processed, given a court hearing, and then either sent to live with family in the United States or deported to another country—Mexico if they are from Mexico.

It was never expected that there would be so many. Conservatives are blaming the situation on Obama, for not just immediately deporting them—though it would be illegal, as well as inhuman, for him to do so. They are not called "Bush's refugee children," though they could be if the issue were just pinning the problem on a conservative rather than treating them humanely—and according to law.

Meanwhile, southern conservatives living near the border are rebelling against treating these refugees humanely and not just deporting them. There are massive conservative-organized protests—people lining the roads waving American flags—shouting racist slogans. Those interviewed in the media say things like *Send them back. They're dirty. They carry diseases. They're criminals. Why is Obama spending our taxpayer dollars to give them clean rooms and clothes and food and medical care? Soon they'll be in our schools. Where are their parents? How could their parents have been so irresponsible to have sent their children here alone? Don't they love their children?*

This is a major humanitarian issue, and it calls for empathy. Parents who love their children don't want to see them maimed or murdered or kidnapped by human traffickers. Many of these children are heroic, somehow traveling over 1,000 miles to get to safety and freedom.

The issue is empathy and respect for these refugees as human beings.

8

The Piketty Insight on the Accelerating Wealth Gap

Systemic causation applies to the economy as well as to global warming, and it has effects every bit as dramatic and crucial. There is an accelerating gap—not just widening but accelerating—between the ultra rich and everyone else. Why? What are the systemic causes and the systemic effects? And is there anything wrong with some people getting that rich and progressively richer over time?

The answers to these questions were sharpened in 2014 by an insight of economist Thomas Piketty and his colleagues—an insight that has not yet become framed in public discourse.

The Piketty insight showed us that our current concept of Rich is not adequate to understand the wealth-gap phenomenon. One needs to also comprehend the notions of Wealth and Proportion of Wealth. Wealth correlates with certain forms of freedom, like the freedom to acquire goods, or to travel, or freedom of access to certain cultural events, and so on. Wealth also correlates with certain forms of power. For example, paying people to do things is a form of power. Contributing significantly to an election campaign can be a form of power, too.

Workers are profit creators, that is, they create wealth for others. They may acquire wealth for their work, but their value to their employers typically lies in the wealth they create for those employers. A natural question is: Of the wealth created by productive work, how much goes to those who do the work and how much goes to others? And by what means? What is the structure of the system that results in the distribution of wealth and the way that distribution changes?

Piketty's *Capital in the Twenty-First Century* is a work of scholarship of the highest order: It changes, or ought to change, not just our understanding of economics, but our understanding of many things. And he has published it just as we need it most.

Here is his basic insight. He studied the history, not just of income, but of wealth. And he observed that there are two fundamentally different kinds of wealth:

- **Productive wealth.** This is wealth generated by work, by producing and selling things or services. The kind of wealth Adam Smith talked about. The prototypical case concerns individuals, for example a baker and a furniture maker. Each makes and sells things, and each needs and buys what the other sells. The baker's income pays the furniture maker, and the furniture maker's income pays the baker. Each works for himself, produces things, gets paid for it, and in a much oversimplified market, each produces wealth for himself and for the other. This is the kind of wealth, productive wealth, measured by the GDP. Piketty calls it "G."
- **Reinvestment wealth.** This is wealth generated by receiving returns on investments and then reinvesting those returns over and over. This kind of wealth grows exponentially, like compound interest. The more have, the more you invest, and the more you invest, the more you have. He calls it "R."

Here's where the concept of proportion comes in. Piketty looks at the proportion of the kinds of wealth, that is, the ratio between R and G over a population. Then he asks, how does it change and why?

His research was done by studying tax records in many countries, dating back to the eighteenth century. What he discovered was that, up until 1913, most wealth was reinvestment wealth. Even during the period of the industrial revolution, which is usually thought of in terms of productive wealth, R was much greater than G. In other words, Piketty showed that the common wisdom is false. Even in capitalist democracies, where individual liberty and the market were supposed to allow for productive wealth through work, it turns out that reinvestment wealth was overwhelming. For instance, in France, a capitalist democracy concerned with égalité, in 1910, 70 percent of the wealth was reinvestment wealth, held by the very wealthy—not productive wealth, distributed over most of the population.

Starting in 1913, there was a major shift. Because of World War I, the Great Depression, and World War II, a significant portion of reinvestment wealth was destroyed. Productive wealth became greater, more G than R. Between 1913 and 1980, most of modern economic theory was developed, whether liberal or conservative. It was primarily based on productive wealth, on GDP—on G, not R.

Then in 1980, something changed—during the Reagan era in America. Reagan greatly cut taxes on the wealthy, started a major attack on unions and thereby on the wages of ordinary workers, cut regulations on business, and so on. Margaret Thatcher did the same in England. And those economic ideas spread. Around 1980, there was a historic shift. R became greater than G again. Reinvestment wealth took over the reins of the modern economies. Being exponential, reinvestment wealth grew exponentially—like compound interest.

In the United States, in 1976 the top 1 percent had 19.9 percent of the wealth. In 2010, the top 1 percent had 35.4 percent of the wealth. In 2010, the top 5 percent had 63 percent of the wealth; and the top 20 percent had 88.9 percent of the wealth. That left the bottom 80 percent with 11.1 percent of the wealth.

That is what the exponential growth of reinvestment wealth leads to. And it gets worse as one goes down the wealth scale, so that six individual members of the Walton family together have more net worth than 41 percent of the families in America (counting families with negative net worth, who are families too).

As the share of the nation's wealth going to the wealthy rises, the share going to everyone else falls. What else falls? The freedom that wealth can buy, the quality of life that wealth can buy, the power that wealth can buy, and the electoral influence that wealth can buy. Technically, we may still have one person, one vote. But the effect of one person on elections has gone way down.

Is this trend reversible? Piketty says yes, but it takes political change.

The Systemic Effects for Politics

Piketty himself is not pessimistic about the fact that R is above G. He points out that political change can bring the runaway accumulation under control, say, via a wealth tax. He also suggests that traditional liberal measures—like raising lower- and middle-class wages, lowering corporate management wages, closing tax loopholes, increasing access to education, and so on—can help bring about such a reversal.

But there are systemic effects that act against a political solution. Greater wealth leads to many things, including:

- **Greater political leverage.** Wealthy people and corporations have great lobbying power with public officials, and it is getting greater all the time.
- **Greater control over public discourse.** Wealthy people and corporations can control public discourse in many ways—by owning media outlets, sponsoring shows, massive advertising, and so on. This control works via the brain. Language and imagery that activate conservative frames will also activate conservative morality—strict father morality in general. As conservative morality gets stronger, progressive morality gets weaker in the brains of the public. This mightily affects what people believe unconsciously as well as consciously, and therefore affects how people vote.
- **Greater control over the rights of others.** Through state control of legislatures, the wealthy can control the voting rights of poorer populations, and state control is cheaper than national control.

What is needed is government payment for elections and serious regulation of political control of the media. But given the present distribution of wealth and the present distribution of strict father morality in the population in the United States, as well as other countries, the necessary political change seems unlikely—unless there are other important changes brought on by progressives willing to build the grounding frames for systemic issues, and to keep the focus on these issues sharp and strong through continual public discourse.

The Effect on Satisfying Productive Work

One of the major systemic effects of the ascendance of reinvestment wealth concerns the nature of productive work itself. It has become less satisfying in many ways. The most obvious is that the productive economic system produces less wealth—it doesn't pay enough for a satisfying life for many of our citizens. It also provides less work—fewer jobs. And the work it does provide is less satisfying.

Satisfying work is about pay and working conditions but also about skilled work that is useful and that people feel good about doing. This work needn't be very high-paying or glamorous, just satisfying. Here are some professions of people I know who have found ways to have satisfying work lives: carpenter, gardener, barber, cheese salesperson, baker, mechanic, office manager, tailor, house painter, chef, tearoom server, schoolteacher, house cleaner, and so on. They are not professionals—not lawyers, doctors, computer scientists, chemists, biologists, or finance professionals, nor musicians, movie actors, or professional athletes. Just folks. They can be well-educated, functioning citizens, good parents. However, fewer and fewer people manage to have such satisfying work lives, and fewer and fewer people are managing to get a real education and function well as community members and parents because of harsher working conditions.

There is a structural reason for this. Remember that companies tend to have two kinds of employees—the assets and the resources. The assets are senior managers and necessary creative people. The resources are people who are interchangeable; who are hired at the lowest possible skill level; the lowest pay and benefit level; and minimally acceptable working conditions such as employment guarantees, pensions, medical care, a pleasant work environment, few if any sick leaves and parental leaves, little choice in work times, few if any raises or bonuses, and so on. This makes for "efficiency," which is defined as the maximization of profit. Workers treated as resources, not assets, are subject to layoffs, buybacks as contract workers, and job loss when outsourcing is more profitable. The corporate movement against unionization not only allows such conditions to occur, but accelerates them.

Computerization and mechanization leads to more and more jobs becoming low-skill, low-pay resource jobs. At the same time, it leads to even greater wealth for corporate managers and investors, since they can either lay off workers or downgrade their skill level and pay scale or outsource to places where labor is cheap.

That tendency is driven by investors' demand for greater and greater reinvestment returns and the drive for managers to become part of the reinvestment wealth class by increasing their wealth. Since corporate managers manage corporate wealth, they can get a greater share of that wealth. Those in a corporation who control where the corporation's money goes can gather to themselves more and more of the corporation's wealth, leaving less for workers who create that wealth.

Are Ordinary Liberal Economic Solutions Inadequate?

Liberals regularly propose measures set within classical liberal economic theory: raising the minimum wage, massive programs for rebuilding infrastructure, better safety nets, early childhood education and better education in general, better health care, and so on. These would help ease the pain on the nonwealthy—and that is a vitally important thing to do. But can liberal economic measures alone overcome runaway accumulation for the rich and runaway loss for the nonrich?

Even to get these easing-the-pain measures, the political climate would have to change radically. And as we have seen, just telling people the Piketty economic facts cannot help, because the Piketty facts will not have real effects without radical framing change.

What's Wrong with Runaway Accumulation?

The major thus-far-unframed effect is that runaway exponential accumulation of wealth share tends to kill off the provision of public resources that makes a satisfying and healthy private life possible. The political effect of runaway wealth is, for example, to cut taxes on the wealthy, taking away funding for the public resources that made that wealth possible in the first place.

Take university education. There are only so many top research universities. A number of them are public. By cutting funding to such "public" universities, these public resources have to raise tuition and other costs and so move away from really being public toward being private. The same goes for education at all levels.

At the same time, real education is being lost. The point of a classic liberal education was manyfold: to develop one's mind and critical faculties in general, to teach about the world so as to open a world of possibilities in life, to provide skills for learning whatever one needs to learn, and to create citizens who contribute to a democratic society. Because of the runaway loss of satisfying work, education has radically changed. More and more students see education as a direct route to either wealth or a satisfying work life—and are therefore getting "educated" for today's jobs, without the intangible but vitally important personal riches of a liberal education. That is educational robbery, because a liberal education opens up possibilities for one's entire life that an orientation toward today's jobs does not— especially when today's jobs may not be there in the future.

The Runaway Loss of Valuable Experiences

If those of great wealth own the beaches, it means all others are deprived of the experience of them. Access for most people is cut off. The loss is a loss of experience. This is true not only of beaches but also of many things that the ultra rich can experience but that the lower-middle class and poor cannot. Excellent schools, pleasant surroundings, summer camps, trips to lovely or interesting places, the ability to visit family, time off from work, expensive art or music events, nice clothes, great food and wine, healthy food, first-rate medical care, major athletic events, world cultures, great cities, film festivals, and on and on. Money buys experiences of personal value— what a lot of life is about. A single wealthy person can only experience so much. The runaway accumulation of wealth for the rich and the runaway loss of wealth for others mean for most people a runaway loss of experiences of personal value—the loss of a meaningful life.

Piketty and Global Warming

The rise of runaway wealth accumulation at the same time as intensifying global warming has created the perfect storm, and these concepts need to be linked in political discourse.

Wealthy corporations and individuals keep reinvesting and getting wealthier. The current framing of global warming in the conservative and often the mainstream media uses both denial and scare tactics like claiming that addressing global warming is too expensive, would ruin the economy, cost massive job layoffs, increase energy dependency, and so on. These are all false claims, as independent studies have shown. But when the wealthy control what appears in the public media, they can control public discourse and public thought mechanisms through the control of language and imagery. And the worse global warming effects get, the greater the pain on the middle and lower classes, while the effects of global warming can be more easily withstood by those with great wealth.

Global warming is the greatest moral issue facing our generation. Accelerating wealth accumulation by the wealthy is a close runner-up. Together, they present a clear and present danger, not just to the United States, but to the world.

Growth

There is a major systemic effect of framing the Piketty insight in terms of inequality alone, and not thinking about how it relates to global warming and other major issues, like the pressure for continued economic growth. Piketty is arguing within traditional economic theory. He estimates that R can in principle (with the right politics) be kept less than G (economic growth measured in terms of GDP) if economic growth G is kept below 2 percent per year.

But growth is compounded and therefore exponential too. Economic growth means population growth, growth in the use of resources, growth in global warming, growth in weather disasters, and growth in the diminishment of the natural world. Over fifty years, even 2 percent growth is huge!

Once one starts talking about global warming, growth itself becomes an issue—population growth, growth in worldwide production for and by that population, growth in food needs, growth in energy needs, growth in natural resource needs, and so on. Fossil fuel use has to be reversed to avoid global warming disasters. An economy based on growth—even as low as 2 percent—in all these areas would not prevent a global warming disaster.

Models for a new "sustainable"—that is, nongrowth—economics are being developed. How does the Piketty insight square with such economic models—if it does at all?

That is a systemic causation question that needs to be asked. For example, the main factor in population growth appears to be women's education and the availability and use of contraception. Women's education is affected by poverty, but every bit as much by religion. Religions like Catholicism and Islamism promote population growth, which makes it harder to control global warming. It isn't just whether R is higher than G.

The Intertwined Systemic Effects

One of the main take-home points of this book is that framing can have massive systemic effects. The absence of adequate framing can have just as massive effects.

Framing the Piketty insight as just about inequality misses most of what we have just discussed. It misses the systemic effects.

Framing is about thought, about understanding at the deepest levels, about circuitry in your brain with strong synapses that last, about changing unconscious, automatic, effortless understanding—in other words, about changing common sense. Frame change itself is a systemic effect. There are a lot of frames to be changed. How can such overall change be effected?

It begins by strengthening the framing for the progressive moral system and for the progressive view of democracy based around empathy and the responsibility flowing from that empathy. In other words, we have to care about others—fellow citizens of the world we have never met and never will meet—and recognize the fact that the private depends on the public.

That in turn depends on another systemic effect—the effect of language and the brain on public discourse, and the failure in universities to teach that effect.

9

Government by Corporation

As we have seen, there is much that is unframed by the general public that needs to be framed, most notably:

- **Runaway wealth to the wealthy.** The wealth share of the most wealthy is growing exponentially, and the wealth share of others is correspondingly declining. In the absence of adequate framing, most people feel the effects but don't comprehend the systemic causes.
- **Runaway climate disasters.** The earth is warming dangerously and quickly, and that warming is systemically causing climate disasters, including extreme cold. Without understanding systemic effects, extreme cold leads to the denial of global warming.
- **Runaway privatization of public resources.** The private depends on the public, but conservatives are drastically cutting funds for public resources while successfully promoting privatization. They say that government doesn't work, and by cutting funds they can make government cease to work. And by cutting government resources for all, they can make democracy cease to work.

But there is an important framing that is beginning to catch on:

- The Constitution applies only to human beings.

Conceptual metaphors have no legal standing. We normally think using thousands of them, but the law does not overtly give them any official role in the law itself. So far as the law is concerned, metaphorical thought, which is ubiquitous, does not exist. But in reality, unconscious conceptual metaphors do exist, they are everywhere, and they have consequences. This disparity between the law and the human brain and mind is unframed—not part of most people's everyday consciousness or discourse. At the same time, a major metaphor has entered into the national consciousness because of certain Supreme Court decisions—that Corporations Are Persons with Constitutional Rights.

Reflexivity enters here. Decisions made by our courts have the power to turn metaphors into facts-on-the-ground, as in this example. Those metaphors taken as facts give rise to further court decisions extending those metaphors.

The power to turn metaphors into facts can be an awesome power—a power with enormous political impact. The Corporations Are Persons metaphor has so great a political impact that it is worth some discussion here. Let's begin, though, by looking at two powerful metaphors that form its roots.

Cognitive scientists who study metaphorical thought have recognized two common metaphors, which we adopt unconsciously and automatically, that are relevant here.

METAPHOR 1: PLURALITIES ARE GROUPS. The Pluralities Are Groups metaphor attributes group properties to separate individuals—whether warranted or not. The result is that the group is perceived as an entity with different properties than the individuals in it. Take a look at the two operative words:

- A **plurality** consists of people, animals, plants, or other things considered separately—ungrouped. A number of people may be riding on the subway, for instance; but aside from being on the same subway, they are not necessarily part of any particular group. They are a plurality—but they need not have common characteristics, goals, or functions.
- A **group** is an entity, conceptualized metaphorically as a container for other entities. The group can, and usually does, have properties, resources, goals, and functions that are separate from the individual entities in the group.

Once we combine the two metaphorically, we being to think about pluralities differently. For example, a club, a church, an association (e.g., the AARP—the American Association of Retired Persons), can have money, a home, legal responsibilities, and liabilities (they can be sued or a lien can be placed against their property) that don't apply to any individual members. Similarly corporations can be sued, while their stockholders can be immune to such lawsuits as individuals.

METAPHOR 2: INSTITUTIONS ARE PERSONS. Ask most people if institutions are people, and they will say no. In fact, our definitions of the two words are quite distinct.

- An **institution** is an abstract entity conceptualized metaphorically as a container for people. An institution is typically defined by its goals, resources, and by various functions, responsibilities, and privileges for whatever people are in the institution. The institution is defined independently of the people who happen to be in it, serving its functions.
- A **person** is a human being. Human beings have goals and resources, and typically have responsibilities, privileges, and carry out functions. Human beings also have properties that institutions do not have: human bodies and brains, feelings and emotions, desires and beliefs, physical functions and needs, as well as social roles and the ability to think and communicate.

But this conceptual metaphor has long been in our brains. We use this metaphorical mode of thought constantly when we are comprehending and discussing institutions, as in: The EPA was disappointed by the court ruling; Major League Baseball wants to wipe out the use of performance-enhancing drugs; Stanford thinks that online courses are a good idea; Berkeley is troubled by rape on campus; Planned Parenthood was disgusted by the recent court decision; and so on.

These two conceptual metaphors—Pluralities Are Groups and Institutions Are Persons—exist in many parts of the world and have for millennia. And in many cases they have been recognized in law.

Roman law recognized certain business and religious groups as institutions with the properties of human beings: goals, resources, functions, responsibilities, and privileges. Today, we still attribute these properties of human beings metaphorically to institutions.

There is a long history of just what properties of human beings are attributed metaphorically to institutions. For example, the Medieval Church saw monasteries as institutions with goals, finances, responsibilities, and privileges—but differing from people in that they lacked a soul. In England, "companies" were institutions given exclusive rights or "charters" to do business for the financial benefit of their shareholders and the British Crown. One of the most successful was the East India Company.

The Massachusetts Bay Colony was founded by the owners of the Massachusetts Bay Company, which had a charter to do business in the New England area. The metaphorical idea that A Government Is a Business came to America in 1623 with the Massachusetts Bay Company, and has been part of American political life ever since.

Before 1819, the commonplace conceptual metaphor that Institutions Are Persons began to be applied to corporations and was limited to such matters as goals, finances, responsibilities, privileges, and so on. But there is a difference between this very common and limited view of corporate personhood and a view that gave corporations constitutional rights! That took the Corporations Are Persons metaphor out of range of normal conceptual metaphor and into the power of the courts.

In 1819, the Supreme Court made a fateful decision in *Trustees of Dartmouth College v. Woodward*. Before the American Revolution, King George III granted a corporate charter to Dartmouth College, giving it land in New Hampshire and giving the trustees the right to administer the college. In 1819, the trustees deposed the college president. The New Hampshire legislature was outraged and passed legislation taking away the college's charter from King George, putting the state in charge, and in effect, making Dartmouth a state college. The trustees brought a case against the state, with Daniel Webster arguing their case passionately before the Supreme Court and Chief Justice John Marshall. The court ruled that, even though all political ties had been severed with King George, the charter still constituted a "contract" with King George, the person. The court held that this contract fell under the Contracts Clause of the Constitution, which forbids a state to pass laws overturning contracts. Though the provision in the Constitution applied to contracts among persons, the court held that it applied to corporations as well. Dartmouth remained private and under the control of the trustees. Meanwhile, a line had been crossed. A constitutional clause granting rights of contract and property to persons now applied to corporations. And an institution, the British monarchy, called metonymically the British Crown, was now taken literally to be a person, King George.

In 1868, the Thirteenth, Fourteenth, and Fifteenth Amendments were passed, making slaves free, giving them equal protection under the law, and guaranteeing them the right to vote. The first clause of the Fourteenth Amendment reads:

> All persons born or naturalized in the United States, and subject to the jurisdiction thereof, are citizens of the United States and of the state wherein they reside. No state shall make or enforce any law which shall abridge the privileges or immunities of citizens of the United States; nor shall any state deprive any person of life, liberty, or property, without due process of law; nor deny to any person within its jurisdiction the equal protection of the laws.

In the years leading up to these amendments, large industries, banks, and railroads took corporate form, and became rich and powerful. Many states set about regulating them and restricting their powers. With the passage of the Fourteenth Amendment intended as protection for former slaves, railroads saw a way around these restrictions.

They assumed the opposite of the Pluralities Are Groups metaphor, which separated the properties of group entities from the individual properties of their members. The new Groups Are

Pluralities metaphor identified the properties of the group with properties of the members. They began arguing that corporations were persons and had as one of their properties the same constitutional protections as persons had. They argued this for nearly twenty years and kept losing. But they got the idea into public discourse, especially among people employed by the railroads.

Then, in 1886, they got a break. The railroads had brought four cases concerning taxes to the Supreme Court, including *Santa Clara County v. Southern Pacific Railroad*. In Santa Clara County, there was a tax provision that allowed persons paying mortgages on their property to deduct the amount of their mortgages from their taxes. Southern Pacific Railroad had a huge mortgage and wanted its mortgage costs deducted from its taxes, which would make a lot of profit for the railroad and cost Santa Clara County a lot of tax money.

The case was argued in the Supreme Court, with Chief Justice Morrison Waite presiding. Waite had earlier been a railroad lawyer. The court recorder was J. C. Bancroft Davis, formerly president of a small railroad.

The precedent of corporate personhood came out of this case. But the precedent did not come from any argument in the case or anything written in favor or against by any of the justices. The precedent came from an oral remark by Chief Justice Waite that was taken down by court reporter J. C. Bancroft Davis. The remark appeared in the headnotes for the case and nowhere else. The headnote quoted Waite as saying:

> The court does not wish to hear argument on the question whether the provision in the Fourteenth Amendment to the Constitution, which forbids any state to deny any person within its jurisdiction the equal protection of the laws, applies to these corporations. We are of the opinion that it does.

That set the precedent, which was then cited in succeeding cases. Bear in mind, as we go through the list of cases, that British common law defined corporations and their shareholders in such a way as to legally separate the properties of the shareholders from the properties of the corporation, so that the shareholders were free of certain liabilities of the corporations—hence the LLC, the limited liability corporation. Each of the following court decisions violated this essential property of corporations, by automatically attributing to corporations certain of the constitutional rights held by shareholders—namely, constitutional rights of due process, free speech (seen as money contributed to political campaigns and spent on the media), freedom of religion, and freedom from unreasonable search and seizure and double jeopardy.

Bear in mind the Mitt Romney quote from the 2012 presidential election: "Corporations are people, my friend. . . . Everything corporations earn eventually goes to people." He neglected to say which people.

In 1889, the court overtly granted due process protections of the Fourteenth Amendment to corporations; in 1893, it was Fifth Amendment protections of double jeopardy; in 1906, Fourth Amendment protections from unreasonable search and seizure; and in 1978, the First Amendment right to make contributions to ballot initiative campaigns.

In the latter case, the metaphor was extended: Though a corporation could not go to the polls and vote, as a "person" it has the right of free speech since its shareholders have the right of free speech. Then a further metaphor: Speech as Money going to campaigns—not for candidates (real people) but for policies affecting corporations. It was a step toward Citizens United.

Interestingly, Citizens United is not directly about corporate personhood and does not depend on that general metaphor. It depends instead on two other metaphors.

- Money Is Speech
- Nonpersons Have the Right to Speech

These two metaphors form a logic: People have a right to as much speech as they want. Since money is speech and nonpersons have the right to speech, it follows that nonpersons have the right to spend as much money as they want on elections.

This 5-to-4 vote by a conservative court was politically motivated. Corporations have much more money to spend on political campaigns than do unions. This ruling gives a huge amount of money to conservatives and hardly any to progressives. Speech, as we have seen, is not just sounding off. If framed and targeted carefully, Citizens United allows conservatives to change the brains of biconceptuals, and win a lot of elections for conservatives and move the country radically to the right.

The Hobby Lobby and Wheaton cases were conservative victories as well. The Hobby Lobby verdict granted First Amendment freedom of religion rights to corporations that are tightly held and operated by a family or small group (more than half controlled by at most five persons). The right given is a *new right*: to ignore a provision of a law applying to a corporation if the small group of individuals owning or controlling the corporation feel that the law violates their religious principles.

This is a new, and very different, metaphorical extension of Corporations Are Persons to First Amendment rights and it opens the floodgates to a huge range of claims to be exempt from provisions of the law on grounds of religious principle as self-defined. In short, it puts corporations above the law. It is a step toward legalizing government by corporations.

This is a radical conservative political decision. Why? Because radical conservatives want to eliminate public resources and public aspects of government—that is, government by laws passed by human legislators. At once, this shifts government from the public to the private sphere and from the human to the nonhuman sphere.

This brings us to another truth unframed in public discourse.

- Corporations govern our lives.

There have been many great innovations made by corporations when they have invested their exponentially accumulated wealth in innovations that improve people's lives—in useful computer technology, telecommunications, pharmaceuticals and medical equipment, transportation—in area after area.

To my knowledge, in all cases, these innovations that have improved our lives were ultimately made possible by public resources: Government-sponsored research and university training made possible computer science, satellites, medical research and training, and so on in case after case. Every great corporate innovation story simply burnishes the truth that the private depends on the public.

But what is left conceptually unframed and therefore unspoken are the negative effects of the runaway accumulation of corporate wealth. Here is a short list.

- **Increasing corporate lobbying and political contributions.** These effects work largely against the public interest on a huge range of issues—even to the extent that corporations write laws introduced by the legislators they contribute to. The Citizens United decision greatly exacerbated this effect.
- **Increasing externalization of costs.** The wealthier corporations get, the more power they have to use their political influence to avoid regulations. As a result, they can pass along to others the costs of doing business—and thus increase profits even more. The fancy name for this is "externalization of costs."

A prime example is the dumping of hazardous waste that taxpayers will pay to clean up—or will suffer with. Consider what happens when fracking companies dump pools of polluted water on the landscape, or tear up the land in the fracking process and then leave it torn up, or inject vast amounts of poisonous chemicals in the porous shale rock next to the water table, thus creating polluted drinking and agricultural water. The burden is shifted away from the private corporations and to the public. The prime example, of course, is of corporations emitting the greenhouse gas pollution that has caused global warming. The costs get dumped onto you—whether you're paying more taxes to mitigate climate change or to clean up after severe storms or you're paying more for vegetables during severe droughts.

But even when you have to spend your time searching a company website or waiting on the phone to talk to a customer service representative, costs are being externalized: Your time is being spent while the company profits by hiring too few people in customer service. Various forms of "self-service" at gas stations, supermarkets, and big box stores are made to sound like conveniences for you, but they are really ways of making you work for the company for free.

- **Increasing costs to consumers due to monopoly ownership.** For example, some Internet providers with no competition may overcharge and provide minimal service, leaving the customer to bear the burden of exorbitant costs and poor service.

- **Limitations of size options by clothing manufacturers.** Many clothing manufacturers will only make sizes that fit the most normally sized people because it is more profitable than providing sizes for the full range of customers.

- **Increasingly unethical business practices.** For example, General Motors sold cars with known defects that caused deaths, while people in the company knew of the dangers and kept silent.

- **Increasing corporate inefficiency.** Anyone who has worked in a large company is familiar with corporate inefficiency (see the Dilbert comic strip). Health insurance companies, for example, have inefficiency costs that are very high compared with Medicare. Those inefficiency costs are transferred to consumers whenever possible.

- **Increasing corporate management pay and the pressure for short-term profits.** When the very rich get exponentially richer and everyone else exponentially loses access to wealth, there is an inevitable pressure for short-term profits. When corporate managers are in charge of managing corporate wealth, there is an incentive for them to acquire exponentially growing wealth.

To a large extent, corporations govern us and run our lives—for their profit, not ours. The list could go on and on.

To a large extent, the runaway expropriation of wealth pointed out by Thomas Piketty is a result of government by corporation.

Piketty points out that a political solution is necessary, but when our politics is governed significantly through lobbying by corporations rather than the public, the possibility of this is greatly reduced.

Conservatives like to rail against "government" as taking away their liberty. But government by corporations probably does far more to take away such "liberty."

Government by corporation is a major unframed reality. It is systemically linked to the runaway accumulation of our wealth by the very wealthy. Because of the systemic effect of runaway personal and corporate wealth on our politics, both are systemically linked to the threat of global warming to the future of our planet, and to the fundamental split in our politics that is systemically threatening democracy in ways that are not obvious, and are therefore also unframed in public discourse.

★★ PART IV ★★

FRAMING: LOOKING BACK A DECADE

10

What's in a Word? Plenty, If It's Marriage

—February 18, 2004, with some updates in 2014—

The original version of this chapter was written over a decade ago, before the major advances in the acceptance of gay marriage. The successful strategy used what was recommended in this chapter in the first edition of this book: the stress on love and commitment and the generalization to everyone, not just gays.

More than half of Americans now support gay marriage. It is legal in nineteen states. But there is still a long way to go in the other thirty-one states. The conservative framing has not changed: It's against the Bible; it threatens the very definition of marriage; it's a lifestyle choice; children will be lured into it; it's all about sex. Conservatives these days repeat the word homosexual, *which contains the word* sex *and the slur* homo.

For example, Texas Governor Rick Perry was reported as saying, "Whether or not you feel compelled to follow a particular lifestyle or not, you have the ability to decide not to do that . . . I may have the genetic coding that I'm inclined to be an alcoholic, but I have the desire not to do that, and I look at the homosexual issue the same way."

For all of that, conservatives are fighting a losing battle against love and commitment, family and community. The younger generation is overwhelmingly accepting.

President Obama entered the presidency not being in favor of gay marriage but being open to "evolving" on the issue. He has now "evolved." The evolution metaphor suggests an adaptation to the changing political context.

What's in a word? Plenty, if the word is *marriage*.

Marriage is central to our culture. Marriage legally confers many hundreds of benefits, but that is only its material aspect. Marriage is an institution, the public expression of lifelong commitment based on love. It is the culmination of a period of seeking a mate, and, for many, the realization of a major goal, often with a buildup of dreams, dates, gossip, anxiety, engagement, a shower, wedding plans, rituals, invitations, a bridal gown, bridesmaids, families coming together, vows, and a honeymoon. Marriage is the beginning of family life, commonly with the expectation of children and grandchildren, family gatherings, in-laws, Little League games, graduations, and all the rest.

Marriage is also understood in terms of dozens of deep and abiding metaphors: a journey through life together, a partnership, a union, a bond, a single object of complementary parts, a haven, a means for growth, a sacrament, a home. Marriage confers a social status—a married couple with new social roles. And for a great many people, marriage legitimizes sex. In short, marriage is a big deal.

In arguing against same-sex marriage, the conservatives are using two powerful ideas: definition and sanctity. We must take them back. We have to fight definition with definition and sanctity with sanctity. As anthropological studies of American marriage have shown, they got the definition wrong. Marriage, as an ideal, is defined as "the realization of love through a lifelong public commitment." Love is sacred in America. So is commitment. There is sanctity in marriage: It is the sanctity of love and commitment.

Like most important concepts, marriage also comes with a variety of prototypical cases: The ideal marriage is happy, lasting, prosperous, and with children, a nice home, and friendships with other married couples. The typical marriage has its ups and downs, its joys and difficulties, typical problems with children and in-laws. The nightmare marriage ends in divorce, due perhaps to incompatibility, abuse, or betrayal. It is a rich concept.

None of the richness we have just discussed requires marriage to be heterosexual—not its definition, its sanctity, its rituals, its family life, its hopes and dreams. The locus of the idea that marriage is heterosexual is in a widespread cultural stereotype.

In evoking this stereotype, language is important. The radical right used to use *gay marriage*; now it's *homosexual marriage*. One reason, I believe, is that marriage evokes the idea of sex, and most

Americans do not favor sex that isn't heterosexual. The stereotype of marriage is heterosexual. *Gay* for the right connotes a wild, deviant, sexually irresponsible lifestyle.

But *gay marriage* is a double-edged sword. President Bush chose not to use the words *gay marriage* in a State of the Union address. I suspect that the omission occurred for a good reason. His position was that *marriage* is defined as being between a man and a woman, and so the term *gay marriage* should be an oxymoron, as meaningless as *gay apple* or *gay telephone*. The more *gay marriage* is used, the more normal the idea of same-sex marriage becomes, and the clearer it becomes that *marriage* is not defined to exclude the very possibility. The grammar is important. *Gay* is grammatically a modifier specifying a kind of marriage. If you understand the expression, then it is not a contradiction in terms and marriage is not "defined" to exclude gays.

Because marriage is central to family life, it has a political dimension. As I discussed earlier, and at greater length in my book *Moral Politics*, conservative and progressive politics are organized around two very different models of married life: a strict father family and a nurturant parent family.

The strict father is moral authority and master of the household, dominating the mother and children and imposing needed discipline. Contemporary conservative politics turns these family values into political values: hierarchical authority, individual discipline, military might. Marriage in the strict father family must be heterosexual marriage: The father is manly, strong, decisive, dominating—a role model for sons, and for daughters a model of a man to look up to.

The nurturant parent model has two equal parents, whose job is to nurture their children and teach their children to nurture others. Nurturance has two dimensions: empathy and responsibility, for oneself and others. Responsibility requires strength and competence. The strong nurturing parent is protective and caring, builds trust and connection, promotes family happiness and fulfillment, fairness, freedom, openness, cooperation, and community development. These are the values of strong progressive politics. Though the stereotype is again heterosexual, there is nothing in the nurturant family model to rule out same-sex marriage.

In a society divided down the middle by these two family models and their politics, we can see why the issue of same-sex marriage is so volatile. What is at stake is more than the material benefits of marriage and the use of the word. At stake are one's identity and most central values. This is not just about same-sex couples. It is about which values will dominate in our society.

When conservatives speak of the "defense of marriage," liberals are baffled. After all, no individual's marriage is being threatened. It's just that more marriages are being allowed. But conservatives see the strict father family, and with it their political values, as under attack. They are right. This is a serious matter for their politics and moral values as a whole. Even civil unions are threatening, since they create families that cannot be traditional strict father families.

Progressives are of two minds. Pragmatic liberals see the issue as one of benefits—inheritance, health care, adoption, and so forth. If that's all that is involved, civil unions should be sufficient—and they certainly were an advance. Civil unions provided equal material protection under the law. Why not leave civil unions to the state and marriage to the churches, as in Vermont, the first state of many to adopt civil unions and later same-sex marriages?

Idealistic progressives saw beyond the material benefits, important as they are. Most gay activists want full-blown marriage, with all its cultural meanings—a public commitment based on love, all the metaphors, all the rituals, joys, heartaches, family experiences—and a sense of normality, on par with all other people. The issue is one of personal freedom: The state should not dictate who should marry whom. It is also a matter of fairness and human dignity. Equality under the law includes social and cultural as well as material benefits. The slogan here is "freedom to marry."

Back in 2004, when the first edition of *Don't Think of an Elephant!* was published, a number of prominent Democrats claimed that marriage was a matter for the church, while the proper role for the state was civil unions and a guarantee of material benefits. This argument has always made little sense to me. The ability of ministers, priests, and rabbis to perform marriage ceremonies is granted by governments, not by religions. And civil marriage is normal and widespread. Besides, it will only satisfy the pragmatic liberals. Idealistic conservatives will see civil unions as tantamount to marriage, and idealistic progressives will see them as falling far short of equal protection.

And what of the constitutional amendment to legally define marriage as between a man and a woman? Conservatives will be for it, and many others with a heterosexual stereotype of marriage may support it. But with nineteen states legalizing gay marriage and a majority of Americans for it, such an amendment is now dead in the water.

Progressives have reclaimed the moral high ground—of the grand American tradition of freedom, fairness, human dignity, and full equality under the law. There is no longer a need to talk about civil unions and just the material benefits. The job of ordinary citizens in the remaining thirty-one states is to reframe the debate, in everything we say and write, in terms of our moral principles.

With the divorce rate for heterosexual marriage skyrocketing, the sanctity of marriage is more important than ever. Talk sanctity. With love and commitment, you have the very definition of the marital ideal—of what marriage is fundamentally about. Any couple willing to fight for a public recognition of their love and a lifetime commitment has sanctity on their side.

We all have to put our ideas out there so that political candidates can readily refer to them. For example, when there is a discussion in your office, church, or other group, there is a simple response for someone who says, "I don't think gays should be able to marry. Do you?" The response is: "I believe in equal rights, period. I don't think the state should be in the business of telling people who they can or can't marry. Marriage is about love and commitment, and denying the right to marry to people in love who want a public lifetime commitment is a violation of human dignity."

The media does not have to accept the right wing's frames, and in state after state they have not. What can a reporter ask besides "Do you support gay marriage?" Try this: "Do you think the government should tell people who they can and can't marry?" Or "Do you think the freedom to marry who you want to is a matter of equal rights under the law?" Or "Do you see marriage as the realization of love in a lifetime commitment?" Or "Does it benefit society when two people who are in love want to make a public lifetime commitment to each other?"

Morally based framing is everybody's job. Especially reporters'.

It has long been right-wing strategy to repeat over and over phrases that evoke their frames and define issues their way. Such repetition makes their language normal, everyday language and their frames normal, everyday ways to think about issues. Reporters have an obligation to notice when they are being taken for a ride, and they should refuse to go along. It is a duty of reporters not to accept such a situation and not to simply use right-wing frames that have come to seem natural. And it is the special duty of reporters to study framing and to learn to see through politically motivated frames, even when those frames have come to be accepted as everyday and commonplace.

11

Metaphors of Terror

—September 16, 2001, edited July 2014—

Our Brains Had to Change

Everything we know is physically instantiated in the neural systems of our brains.

What we knew before September 11 about America, Manhattan, the World Trade Center, air travel, and the Pentagon were intimately tied up with our identities and with a vast amount of what we took for granted about everyday life. It was all there physically in our neural synapses. Manhattan: the gateway to America for generations of immigrants—the chance to live free of war, pogroms, religious and political oppression!

The Manhattan skyline had meaning in my life, even more than I knew. When I thought of it, I thought of my mother. Born in Poland, she arrived as an infant; grew up in Manhattan; worked in factories for twenty-five years; and had family, friends, a life, a child. She didn't die in concentration camps. She didn't fear for her life. For her, America was not all that she might have wanted it to be, but it was plenty.

I grew up in Bayonne, New Jersey, across the bay from that skyline. The World Trade Center wasn't there then, but over the years, as the major feature of the skyline, it became for me, as for others, the symbol of New York—not only the business center of America but also the cultural center and the communications center. As such, it became a symbol for America itself, a symbol for what it meant to be able to go about your everyday life free of oppression, and just able to live and do your job, whether as a secretary or an artist, a manager or a fireman, a salesman or a teacher or a TV star. I wasn't consciously aware of it, but those images were intimately tied to my identity, both as an individual and as an American. And all that and so much more were there physically as part of my brain on the morning of September 11, 2001.

The devastation that hit those towers that morning hit me. Buildings are metaphorically people. We see features—eyes, nose, and mouth—in their windows. I now realize that the image of the plane going into South Tower was for me an image of a bullet going through someone's head, the flames pouring from the other side like blood spurting out. It was an assassination. The tower falling was a body falling. The bodies falling were me, relatives, friends. Strangers who had smiled as they had passed me on the street screamed as they fell past me. The image afterward was hell: ashes, smoke and steam rising, the building skeleton, darkness, suffering, death.

The people who attacked the towers got into my brain, even three thousand miles away. All those symbols were connected to more of my identity than I could have realized. To make sense of this, my very brain had to change. And change it did, painfully. Day and night. By day the consequences flooded my mind; by night the images had me breathing heavily, nightmares keeping me awake. Those symbols lived in the emotional centers of my brain. As their meanings changed, I felt emotional pain.

It was not just me. It was everyone in this country, and many in other countries. The assassins managed not only to kill thousands of people but also to reach in and change the brains of people all over America.

It is remarkable to know that two hundred million of my countrymen feel as wrenched as I do.

The Power of the Images

As a metaphor analyst, I want to begin with the power of the images and where that power comes from.

There are a number of metaphors for buildings. A common visual metaphor is that buildings are heads, with windows as eyes. The metaphor is dormant, there in our brains, waiting to be awakened. The image of the plane going into South Tower of the World Trade Center activated it. The tower became a head, with windows as eyes, the edge of the tower the temple. The plane going through it became a bullet going through someone's head, the flames pouring from the other side the blood spurting out.

Metaphorically, tall buildings are people standing erect. As each tower fell, it became a body falling. We are not consciously aware of the metaphorical images, but they are part of the power and the horror we experience when we see them.

Each of us, in the premotor cortex of our brains, has what are called mirror neurons connected to visual areas. Such neurons fire either when we perform an action or when we see the same action performed by someone else. There are connections from that part of the brain to the emotional centers. Such neural circuits are believed to be the basis of empathy.

This works literally: When we see a plane coming toward the building and imagine people in the building, we feel the plane coming toward us; when we see the building toppling toward others, we feel the building toppling toward us. It also works metaphorically: If we see the plane going through the building, and unconsciously we evoke the metaphor of the building as a head with the plane going through its temple, then we sense—unconsciously but powerfully—being shot through the temple. If we evoke the metaphor of the building as a person and see the building fall to the ground in pieces, then we sense—again unconsciously but powerfully—that we are falling to the ground in pieces. Our systems of metaphorical thought, interacting with our mirror neuron systems, turn external literal horrors into felt metaphorical horrors.

Here are some other metaphorical and symbolic effects:

- **Control is up.** You have control over the situation; you're on top of things. This has always been an important basis of towers as symbols of power. In this case, the toppling of the towers meant loss of control, loss of power.
- **Phallic imagery.** Towers are symbols of phallic power, and their collapse reinforces the idea of loss of power. Another kind of phallic imagery was more central here: the planes penetrating the towers with a plume of heat, and the Pentagon, a

vaginal image from the air, penetrated by the plane as missile. These phallic interpretations came from women who felt violated both by the attack and the images on TV.

- **A society is a building.** A society can have a "foundation," which may or may not be solid, and it can "crumble" and "fall." The World Trade Center was symbolic of American society. When it crumbled and fell, the threat was to more than a building.
- **Standing.** We think metaphorically of things that perpetuate over time as "standing." During the Gulf War, George H. W. Bush kept saying, "This will not stand," meaning that the situation would not be perpetuated over time. The World Trade Center was built to last ten thousand years. When it crumbled, it metaphorically raised the question of whether American power and American society would last. And that was why it was attacked.
- **Building as temple.** Here we had the destruction of the temple of capitalist commerce, which lies at the heart of our society.
- **Our minds play tricks on us.** The image of the Manhattan skyline became unbalanced. We were used to seeing it with the towers there. Our minds imposed our old image of the towers, and the sight of them gone gave one the illusion of imbalance, as if Manhattan were sinking. Given the symbolism of Manhattan as the promise of America, it appeared metaphorically as if that promise were sinking.

    The Freedom Tower now stands at 1 World Trade Center. It's not as distinctive and its meaning is not the same. It does not represent the stability of normal life in America.
- **Hell.** We had the persistent image, day after day, of the charred and smoking remains: hell.

The World Trade Center was a potent symbol, tied into our understanding of our country and ourselves in myriad ways. All of what we know is physically embodied in our brains. To incorporate the new knowledge requires a physical change in the synapses of our brains, a physical reshaping of our neural system.

The physical violence was not only in New York and Washington. Physical changes—violent ones—have been made to the brains of all Americans.

How the Administration Framed the Event

The Bush administration's framings and reframings and its search for metaphors should be noted. The initial framing was as a crime with victims, and perpetrators to be "brought to justice" and "punished." The crime frame entails law, courts, lawyers, trials, sentencing, appeals, and so on. It was hours before crime changed to war, with casualties, enemies, military action, war powers, and so on.

Donald Rumsfeld and other Bush administration officials pointed out that this situation did not fit our understanding of a war. There were enemies and casualties all right, but no enemy army, no regiments, no tanks, no ships, no air force, no battlefields, no strategic targets, and no clear victory. The war frame just didn't fit. Colin Powell had always argued that no troops should be committed without specific objectives, a clear and achievable definition of victory, and a clear exit strategy, and that open-ended commitments should not be used. But he also pointed out that none of these was present in this "war."

Because the concept of war didn't fit, there was a frantic search for metaphors. First, Bush called the terrorists cowards—but this didn't seem to work too well for martyrs who willingly sacrificed their lives for their moral and religious ideals. Then he spoke of "smoking them out of their holes," as if they were rodents, and Rumsfeld spoke of "drying up the swamp they live in," as if they were snakes or lowly swamp creatures. The conceptual metaphors here were moral is up, immoral is down (they are lowly), and immoral people are animals (that live close to, or in, the ground).

Bush speechwriter David Frum created the phrase *Axis of Evil* which was used in Bush's 2002 State of the Union address to refer to Iran, Iraq, and North Korea and was used over and over by the Bush administration in justifying the war in Iraq. *Axis* was a reference to the enemy Axis powers of World War II—Germany, Italy, and Japan—which spanned the Western and Eastern hemispheres and represented the global distribution of America's deadly enemies. Lumping Iraq with Iran and North Korea suggested that Iraq was developing nuclear weapons (the nonexistent "weapons of mass destruction") and gave a pretext for invading Iraq. *Axis*, because it included Japan, evoked the "sneak attack" on Pearl Harbor, and symbolically equated the September 11 attack with Pearl Harbor, again as a justification for going to war. On the assumption that America contains the essence of morality and democracy—the shining city on the hill—anyone attacking America was evil. And what happened on September 11 certainly felt evil.

The use of the word *evil* in the Bush administration's discourse worked in the following way. In conservative, strict father morality (see *Moral Politics*, chapter 5), evil is a palpable thing, a force in the world. To stand up to evil you have to be morally strong. If you're weak, you let evil triumph, so that weakness in itself is a form of evil, as is promoting weakness. Evil is inherent, an essential trait, that determines how you will act in the world. Evil people do evil things. No further explanation is necessary. There can be no social causes of evil, no religious rationale for evil, no reasons or arguments for evil. The enemy of evil is good. If our enemy is evil, we are inherently good. Good is our essential nature, and what we do in the battle against evil is good. Good and evil are locked in a battle, which is conceptualized metaphorically as a physical fight in which the stronger wins. Only superior strength can defeat evil, and only a show of strength can keep evil at bay. Not to show overwhelming strength is immoral, since it will induce evildoers to perform more evil deeds, because they'll think they can get away with it. To oppose a show of superior strength is therefore immoral. Nothing is more important in the battle of good against evil, and if some innocent noncombatants get in the way and get hurt, it is a shame, but it is to be expected and nothing can be done about it. Indeed, performing lesser evils in the name of good is justified—"lesser evils" like curtailing individual liberties, sanctioning political assassinations, overthrowing governments, torturing, hiring criminals, and creating "collateral damage."

Then there is the basic security metaphor, security as containment—keeping the evildoers out. Secure our borders, keep them and their weapons out of our airports, have marshals on the planes. Most security experts say that there is no sure way to keep terrorists out or to deny them the use of some weapon or other; a determined, well-financed terrorist organization can penetrate any security system. Or they can choose other targets—say, oil tankers.

Yet the Security as Containment metaphor is powerful. It is what lies behind the missile shield proposal. Rationality might say that the September 11 attacks showed the missile shield is pointless. But it strengthened the use of the Security as Containment metaphor. As soon as you say "national security," the Security as Containment metaphor will be activated, and with it a missile shield.

The Conservative Advantage

The reaction of the Bush administration was just what you would expect a conservative reaction would be: pure strict father morality. There is evil loose in the world. We must show our strength and wipe it out. Retribution and vengeance are called for. If there are casualties or collateral damage, so be it.

The reaction from liberals and progressives was far different: *Justice is called for, not vengeance.* Understanding and restraint are what is needed. The model for our actions should be the rescue workers and doctors—the healers—not the bombers.

We should not be like *them*. We should not take innocent lives in bringing the perpetrators to justice. Massive bombing of Afghanistan and Iraq—with the killing of innocents—will show that we are no better.

But the Bush administration's conservative message dominated the media. The event was framed in their terms. As Newt Gingrich put it on the Fox network, "Retribution is justice."

It is vital to understand the history of this framing now, because it is coming up again, with the conservative attack on Obama's call for the use of "soft power"—diplomacy and economic leverage—and the conservatives' call for a military buildup and intervention in the world's trouble spots.

I have been reminded, over and over, of Gandhi's words: "Be the change you wish to see in the world." The words apply to governments as well as to individuals.

Causes

There are (at least) three kinds of causes of radical Islamic terrorism:

- Worldview: the religious rationale
- Social and political conditions: cultures of despair
- Means: the enabling conditions

The Bush administration discussed only the third: the means that enable attacks to be carried out. These include leadership (for example, bin Laden), host countries, training facilities and bases, financial backing, cell organization, information networks, and so on. These do not include the first and second on the list.

Worldview: The Religious Rationale

After the bombing occurred, the question that kept being asked in the media was, "Why do they hate us so much?"

It is important at the outset to separate moderate-to-liberal Islam from radical Islamic fundamentalists, who do not represent most Muslims.

Radical Islamic fundamentalists hate our culture. They have a worldview that is incompatible with the way that Americans—and other Westerners—live their lives.

One part of this worldview concerns women, who are to hide their bodies, should have no right to property, and so on. Western sexuality, mores, music, and women's equality all violate their values, and the ubiquity of American cultural products, like movies and music, throughout the world offends them.

A second part concerns theocracy: They believe that governments should be run by clerics according to strict Islamic law.

A third concerns holy sites, like those in Jerusalem, which they believe should be under Islamic political and military control.

A fourth concerns the commercial and military incursions by Westerners on Islamic soil, which they liken to the invasion of the hated crusaders. The way they see it, our culture spits in the face of theirs.

A fifth concerns jihad—a holy war to protect and defend the faith. A sixth is the idea of a martyr, a man willing to sacrifice himself for the cause. His reward is eternal glory—an eternity in heaven surrounded by willing young virgins. In some cases there is a promise that his family will be taken care of by the community.

Social and Political Conditions: Cultures of Despair

Most Islamic would-be martyrs not only share these beliefs but also have grown up in a culture of despair; they have nothing to lose. Eliminate such poverty and you eliminate the breeding ground for most terrorists—though the September 11 terrorists were relatively well-to-do. When the Bush administration spoke of eliminating terror, it did not appear to be talking about eliminating cultures of despair and the social conditions that lead one to want to give up life to martyrdom.

Princeton Lyman of the Aspen Institute made an important proposal—that the worldwide antiterrorist coalition being formed should address the causal real-world conditions. Country by country, the conditions (both material and political) leading to despair need to be addressed, with a worldwide commitment to ending them. It should be done because it is a necessary part of addressing the causes of terrorism—and because it is right! It never happened.

What about the first cause—the radical Islamic worldview itself? Military action won't change it. Social action won't change it. Worldviews live in the minds of people. How can one change those minds—and if not present minds, then future minds? The West cannot! Those minds can only be changed by moderate and liberal Muslims—clerics, teachers, elders, respected community members. There is no shortage of them. I doubt that they are well organized, but the world needs them to be well organized and effective. It is vital that moderate and liberal Muslims form a unified voice against hate and, with it, terror. Remember that *Taliban* means "student." Those who teach hate in Islamic schools must be replaced—and we in the West cannot replace them. This can only be done by an organized moderate, nonviolent Islam. The West can make the suggestion and offer extensive resources, but we alone are powerless to carry it out. We depend on the goodwill and courage of moderate Islamic leaders. To gain it, we must show our goodwill by beginning in a serious way to address the social and political conditions that lead to despair.

Thinking of the enemy as evil means not taking the primary causes seriously.

Public Discourse

The Honorable Barbara Lee (D-Calif.), who I am proud to acknowledge as my representative in Congress, in casting the lone vote against giving President Bush full Congressional approval for carrying out his War on Terrorism as he saw fit, said the following:

> I am convinced that military action will not prevent further acts of international terrorism against the United States. This is a very complex and complicated matter.
>
> . . . However difficult this vote may be, some of us must urge the use of restraint. Our country is in a state of mourning. Some of us must say, let us step back for a moment. Let us just pause for a minute and think through the implications of our actions today so that this does not spiral out of control.
>
> I have agonized over this vote, but I came to grips with it today and I came to grips with opposing this resolution during the very painful yet very beautiful memorial service. As a member of the clergy so eloquently said, "As we act, let us not become the evil that we deplore."

I agree. But what is striking to me as a linguist is the use of negatives in the statement: "not prevent," "restraint" (inherently negative), "not spiral out of control," "not become the evil that we deplore." A petition was circulated calling for "justice *without* vengeance." *Without* has another implicit negative. It is not that these negative statements are wrong (three negatives!). What is needed is a *positive* form of discourse.

There is one.

The central concept is that of responsibility, which is at the heart of progressive/liberal morality. (See *Moral Politics*.) Progressive/liberal morality begins with empathy, the ability to understand others and feel what they feel. That is presupposed in responsibility—responsibility for oneself, for protection, for the care of those who need care, and for the community. Those were the values that we saw in action among the rescue workers in New York right after the September 11 attack.

Responsibility requires competence and effectiveness. If you are to deal responsibly with terrorism, you must deal effectively with all its causes: religious, social, and enabling. The enabling causes must be dealt with effectively. Bombing innocent civilians and harming them by destroying their country's domestic infrastructure will be counterproductive—as well as immoral. Responsibility requires care in the place of blundering, overwhelming force.

Massive bombing would be irresponsible. Failure to address the religious and social causes would be irresponsible. The responsible response begins with joint international action to address all three: the social and political conditions and the religious worldview and the means, with all due care.

Foreign Policy

At a time when terrorist threats come from groups of individuals rather than states, when wars occur within nations, when "free markets" exist without freedom, when overpopulation threatens stability and disastrous global warming, when intolerant cultures limit freedom and promote violence, when transnational corporations act like oppressive governments, and when the oil economy threatens the planet's future, the central problems in today's world cannot be solved by state-level approaches.

The state-level part of the answer is to recognize global interdependence and focus foreign policy on diplomacy, alliances, international institutions, and strong defensive and peacekeeping forces, with war as a last resort.

But what is needed even more is a moral foreign policy, one that realizes that America can only be a better America if the world is a better world. America must become a moral leader using fundamental human values: caring and responsibility carried out with strength to respond to the world's problems.

In a values-based foreign policy, issues that were not previously seen as part of foreign policy become central. Women's education is the best way to alleviate overpopulation and promote development. Renewable energy could make the world oil-independent. Food, water, health, ecology, and corporate reform are foreign policy issues, as are rights: rights of women, children, workers, prisoners, refugees, and political minorities.

These issues were previously left to international advocacy groups, and many have done excellent work. But these issues need an integrated approach that requires a foreign policy that is serious about addressing them.

The Obama administration is making moves in this direction, with the understanding that these are foreign policy issues. The president is being attacked for it. And the media has not yet recognized them as foreign policy issues. Why?

The metaphors that foreign policy experts have traditionally used to define what foreign policy is have ruled out these important concerns. The metaphors involve self-interest (for example, the rational actor model), stability (a physics metaphor), industrialization (unindustrialized nations are "underdeveloped"), growth (our current economics depends on growth—of markets and access to cheap labor and abundant cheap resources) and trade (freedom is free trade).

There is an alternative way of thinking about foreign policy under which all these humanistic issues would become a natural part of what foreign policy is about. The premise is that when international relations work smoothly, it is because certain moral norms of the international community are being followed. This mostly goes unnoticed, since those norms are commonly followed. We notice problems when those norms are breached. Given this, it makes sense that foreign policy should be centered around those norms.

The moral norms I suggest come out of what in *Moral Politics* I called nurturant morality. It is a view of ethical behavior that centers on empathy and responsibility (for yourself and others needing your help). Many things follow from these central principles: fairness, minimal violence (for example, justice without vengeance), an ethic of care, protection of those needing it, a recognition of interdependence, cooperation for the common good, the building of community, mutual respect, and so on. When applied to foreign policy, nurturant moral norms would lead the American government to uphold the Anti-Ballistic Missile (ABM) Treaty, sign and uphold international environmental accords, engage in a form of globalization governed by an ethics of care—and it would automatically make all the concerns listed above (such as the environment and women's rights) part of our foreign policy.

This, of course, implies (1) multilateralism, (2) interdependence, and (3) international cooperation. But these three principles, without nurturant norms, could equally well apply to a radically conservative foreign policy. Bush's foreign policy was one of self-interest ("what's in the best interest of the United States")—if not outright hegemony (the Cheney/Rumsfeld position). The Democratic leaders incorrectly criticized Bush for being isolationist and unilateralist on issues like the Kyoto accords and the ABM Treaty. He was neither isolationist nor unilateralist. He was just following his stated policy of self-interest, using strict father morality as his guide.

Imagine if Bush had happened to receive the full support of France, Germany, and the United Nations when he announced his policy. Then he would have been called an internationalist and multilateralist. When it was in America's interest (as he saw it), he would work with those nations willing to go along, "the coalition of the willing." Whether Bush looked like a multilateralist depended on who was willing. Self-interest crosses the boundaries between unilateralism and multilateralism. The Bush foreign policy was one of unyielding self-interest.

There is, interestingly, an apparent overlap between the nurturant norms policy and an idealistic vision of the war started by the Bush administration. The overlap is, simply, that it is a moral norm to refuse to engage in or support terrorism. From this perspective, it looked, at the onset, like left and right were united. It was an illusion.

In nurturant norms policy, antiterrorism arises from another moral norm: *Violence against innocent parties is immoral.*

Within a year of the end of the Gulf War, the CIA reported that about a million Iraqi civilians had died from the effects of the war and the embargo—many from disease and malnutrition due to the US destruction of water treatment plants, hospitals, electric generation plants, and so on, together with the inability to get food and medical supplies. Many more innocents have since died from the effects of the war.

In conservative morality, there is a fight between good and evil, in which "lesser evils" are tolerated and even seen as necessary and expected.

The argument that killing innocent civilians in retaliation would make us as bad as them may work for progressives, but not for conservatives.

Be the change you wish to see in the world! If the United States wants terror to end, the United States must end its own contribution to terror.

The foreign policy of moral norms—of diplomacy and economic leverage, minimizing the use of force—is the only sane and humane foreign policy.

Domestic Policy

After September 11, I had a rational fear: a fear that the attack had given the Bush administration a free hand in pursuing a conservative domestic agenda. It could not be said by the media at the time, but it was true. No taxes from the wealthy were raised to pay for the war. Indeed, they were cut! The social security surplus was taken from the "lockbox" to pay for the war, with every Democrat in Congress except one voting for it.

It was supposed to cost forty billion dollars. It actually cost three trillion so far—and counting, if you count the continuing cost of treating veterans injured in the war and of propping up governments

in Iraq and Afghanistan. It ruined the economy of our country—
our educational system, our infrastructure—and it has taken away a
vast range of needed public resources. The poor and middle class got
poorer, but the rich got richer. And the earth got warmer. And the
conservative movement grew.

It was another lesson in systemic causation. Foreign policy and
domestic policy are inextricably linked. Guns made for war will be
sold at gun shows and used to kill children. Drones and computer
technology developed for the surveillance of enemies abroad will be
used for surveillance of civilians at home. And money spent on war
abroad will be drained from public resources at home.

12

Metaphors That Kill

—March 18, 2003, edited July 2014—

The 2003 version of this chapter appeared just before the start of the Iraq War. It is reprinted here to provide a sense of what the study of framing brought to an understanding of that war before it happened.

Metaphors can kill.

That's how I began a piece on the Gulf War back in 1990, just before the war began. (See georgelakoff.files.wordpress.com/2011/04 /metaphor-and-war-the-metaphor-system-used-to-justify-war-in-the -gulf-lakoff-1991.pdf.) Many of those metaphorical ideas are back, but within a very different and more dangerous context. Since the Iraq War is due to start any day, perhaps even tomorrow, it might be useful to take a look before the action begins at the metaphorical ideas being used to justify the Iraq War.

One of the central metaphors in our foreign policy is that a nation is a person. It is used hundreds of times a day, every time the nation of Iraq is conceptualized in terms of a single person, Saddam Hussein. The war, we are told, is not being waged against the Iraqi people, but only against this one person. Ordinary American citizens are using this metaphor when they say things like "Saddam is a tyrant. He must be stopped." What the metaphor hides, of course, is that the three thousand bombs to be dropped in the first two days will not be dropped on that one person. They will kill many thousands of people hidden by the metaphor, people that we are, according to the metaphor, not going to war against.

The Nation as a Person metaphor is pervasive, powerful, and part of an elaborate metaphor system. It is part of an international community metaphor in which there are friendly nations, hostile nations, rogue states, and so on. This metaphor comes with a notion of the national interest: Just as it is in the interest of a person to be healthy and strong, so it is in the interest of a nation-person to be economically healthy and militarily strong. That is what is meant by the "national interest."

In the international community, peopled by nation-persons, there are nation-adults and nation-children, with maturity metaphorically understood as industrialization. The children are the "developing" nations of the third world, in the process of industrializing, who need to be taught how to develop properly and must be disciplined (say, by the International Monetary Fund) when they fail to follow instructions. "Backward" nations are those that are "underdeveloped." Iraq, despite being the cradle of civilization, is seen via this metaphor as a kind of defiant, armed teenage hoodlum who refuses to abide by the rules and must be taught a lesson.

The international relations community adds to the Nation as a Person metaphor what is called the rational actor model. The idea here is that it is irrational to act against your interests, and that nations act as if they were rational actors—individual people trying to maximize their gains and assets and minimize their costs and losses. In the Gulf War, the metaphor was applied so that a country's "assets" included its soldiers, matériel, and money. Since the United States lost few of those "assets" in the Gulf War, the war was reported, just afterward in the *New York Times* business section, as having been a "bargain." Because Iraqi civilians were not our assets they could not be counted among the "losses," and so there was no careful public accounting of civilian lives lost, people maimed, and children starved or made seriously ill by the war or the sanctions that followed it. Estimates vary from half a million to a million or more. However, public relations was seen to be a US asset: Excessive slaughter reported in the press would be bad PR, a possible loss. These metaphors are with us again. A short war with few US casualties would minimize costs. But the longer it goes on, the more Iraqi resistance and the more US casualties, the less the United States would appear invulnerable and the more the war would appear as a war against the Iraqi people. That would be a high "cost."

According to the rational actor model, countries act naturally in their own best interests—preserving their assets, that is, their own populations, their infrastructures, their wealth, their weaponry, and so on. That is what the United States did in the Gulf War and what it is doing now. But Saddam Hussein, in the Gulf War, did not fit our government's rational actor model. He had goals like preserving his power in Iraq and being an Arab hero just for standing up to the Great Satan. Though such goals might have their own rationality, they are "irrational" from the model's perspective.

One of the most frequent uses of the Nation as a Person metaphor comes in the almost daily attempts to justify the war metaphorically as a "just war." The basic idea of a just war uses the Nation as a Person metaphor, plus two narratives that have the structure of classical fairy tales: the self-defense story and the rescue story.

In each story there is a hero, a crime, a victim, and a villain. In the self-defense story the hero and the victim are the same. In both stories the villain is inherently evil and irrational: The hero can't reason with the villain; he has to fight him and defeat or kill him. In both, the victim must be innocent and beyond reproach. In both, there is an initial crime by the villain, and the hero balances the moral books by defeating him. If all the parties are nation-persons, then self-defense and rescue stories become forms of a just war for the hero-nation.

In the Gulf War, George H. W. Bush tried out a self-defense story: Saddam was "threatening our oil lifeline." The American people didn't buy it. Then he found a winning story, a rescue story: the "rape" of Kuwait. It sold well, and is still the most popular account of that war.

In the Iraq War, George W. Bush is pushing different versions of the same two story types, and this explains a great deal of what is going on in the American press and in speeches by Bush and Powell. If they can show that Saddam Hussein equals Al-Qaeda—that he is helping or harboring Al-Qaeda—then they can make a case for the self-defense scenario, and hence for a just war. Or if weapons of mass destruction ready to be deployed are found, the self-defense scenario can be justified in another way.

Indeed, despite the lack of any positive evidence and the fact that the secular Saddam and the fundamentalist bin Laden despise each other, the Bush administration has managed to convince 40 percent of the American public of the link just by asserting it. The administration has told its soldiers the same thing, and so our military personnel see themselves as going to Iraq in defense of their country. In the rescue scenario the victims are (1) the Iraqi people and (2) Saddam's neighbors, whom he has not attacked but is seen as threatening. That is why Bush and Powell keep on listing Saddam's crimes against the Iraqi people and the weapons he could use to harm his neighbors. Again, most of the American people have accepted the idea that the Iraq War is a rescue of the Iraqi people and a safeguarding of neighboring countries. Actually, the war threatens the safety and well-being of the Iraqi people.

And why such enmity toward France and Germany? Via the Nation as a Person metaphor, they are supposed to be our "friends," and friends are supposed to be supportive and jump in and help us when we need help. Friends are supposed to be loyal. That makes France and Germany fair-weather friends! Not there when you need them.

This is how the war is being framed for the American people by the administration and the media. Millions of people around the world can see that the metaphors and fairy tales don't fit the current situation, that the Iraq War does not qualify as a just war—a "legal" war. But if you accept all these metaphors, as Americans have been led to do by the administration, the press, and the lack of an effective Democratic opposition, then the Iraq War would indeed seem like a just war.

But surely most Americans have been exposed to the facts—the lack of a credible link between Saddam and Al-Qaeda, no WMDs (weapons of mass destruction) found, and the idea that large numbers of innocent Iraqi civilians will be killed or maimed by our bombs. Why don't they reach the rational conclusion?

One of the fundamental findings of cognitive science is that people think in terms of frames and metaphors—conceptual structures like those we have been describing. The frames are in the synapses of our brains, physically present in the form of neural circuitry. When the facts don't fit the frames, the frames are kept and the facts ignored.

It is a common folk theory of progressives that "the facts will set you free." If only you can get all the facts out there in the public eye, then every rational person will reach the right conclusion. It is a vain hope. Human brains just don't work that way. Framing matters. Frames once entrenched are hard to dispel.

In the Gulf War, Colin Powell began the testimony before Congress. He explained the rational actor model to Congress and gave a brief exposition of the views on war of Clausewitz, the Prussian general: War is business and politics carried out by other means. Nations naturally seek their self-interest, and when necessary they use military force in the service of their self-interest. This is both natural and legitimate.

To the Bush administration, this war furthers our self-interest in controlling the flow of oil from the world's second-largest known reserve, and in being in the position to control the flow of oil from central Asia. This would guarantee energy domination over a significant part of the world. The United States could control oil sales around the world. And in the absence of alternative fuel development, whoever controls the worldwide distribution of oil controls politics and economics.

My 1990 paper did not stop the Gulf War. This paper will not stop the Iraq War. So why bother?

I think it is crucially important to understand the cognitive dimensions of politics—especially when most of our conceptual framing is unconscious and we may not be aware of our own metaphorical thought. I have been referred to as a "cognitive activist," and I think the label fits me well. As a professor I do analyses of linguistic and conceptual issues in politics, and I do them as accurately as I can. But that analytic act is a political act. Awareness matters. Being able to articulate what is going on can change what is going on—at least in the long run.

★★ PART V ★★

FROM THEORY TO ACTION

13

What Conservatives Want

Liberals tend not to understand conservatives, and their confusion is showing. On the one hand liberals see conservatives in disarray and react with glee at the fragmentation: the Tea Party vs. Libertarians vs. Neocons vs. Wall Street. Eric Cantor, the Republican Majority Leader, brought down by a Tea Party unknown. John Boehner unable to control his majority in the House. Republican primary challenges everywhere.

On the other hand, liberals are scared stiff of the Koch brothers and other wealthy Republicans bankrolling Republican candidates at every level all over the country. They are scared of a Republican takeover. And they should be.

Which is it?

There are real splits, disputes, dislikes, even hatreds among conservatives. Is it tearing conservatism apart?

Many say yes. The tearing-apart theory is easy to understand and constantly discussed.

On the other hand, it is also possible that the divisions form a system that welds the diverse parts together. And this may be making conservatives stronger at the nexus points, where views are shared, not weaker.

The welding-together theory has not been considered, but it is quite possible that this is what is happening among conservatives, at the systems level. Consider the uniformity of opinion among conservatives on everything from Obamacare to abortion to the Supreme Court's Hobby Lobby decision. It is strong, and there are many instances where conservatives of all stripes are fervently against all liberals and all liberal positions. Where they have found common ground fits consistently within their overriding moral framework, and where they have found common ground, their resolve strengthens.

And if it is true that the divisions among conservatives make them stronger, not weaker, then progressives had better be aware of it.

No matter what progressives believe about conservative fragmentation, though, they need to understand who conservatives are and what they want, both group by group and on the whole.

At the heart of conservatism is strict father morality, as we have seen. But strict father morality has complexities and natural variations. What liberals don't see is that the diversity can give conservatism as a whole considerable strength.

Different versions of conservatisms are defined by particular domains of interest. Strict father morality applies to all the domains—individual liberty and self-interest, world power, business, and society. These domains of interest characterize libertarian, neocon, financial, and Tea Party conservatives.

Domain of Interest	Type of Conservative
Individual liberty	Libertarian
World Power	Neocon
Business	Wall Street
Society and Religion	Tea Party

They have the same general strict father morality, but apply it to the domains that they care about most in different ways. The split is not in the moral theory, but in the domains of interest. With complementary differences, they stand together as one.

A focus on unimpeded pursuit of self-interest—and with it, extreme limits on state power over the individual—defines the libertarian strain of right-wing thought.

Neocons believe in the unbridled use of power (including state power) to extend the reign of strict father values and ideas into every domain, domestic and especially international. They are concerned with global financial and military power, and the use of power at home. They sometimes run up against libertarians, who object to the use of governmental power and to global involvements that require the buildup and use of state power.

Wall Street conservatives are primarily concerned with the acquisition of wealth via the corporate world. They include CEOs and upper management of wealthy corporations, investment bankers, venture capitalists, private asset managers, hedge fund managers, and anyone whose income primarily comes from investments. Such conservatives have many political concerns: tax policy, economic treaties, import-export policy, protection of foreign investments, government contracts, access to minerals on government lands, protection of patents and copyrights, property rights versus environmental rights, energy supplies, control of markets, privatization of public resources, and so on. They tend to work through lobbyists, advertising, and control of the media and public discourse.

Finally, there are Tea Party conservatives—social and religious culture warriors, who want to act aggressively on every front in the culture war against liberals and progressives.

On the whole, the right wing is attempting to impose a strict father ideology on America and, ultimately, the rest of the world. Although the details vary with conservative areas of concern, there are general tendencies. Many progressives underestimate just how radical an ideology this is.

Here is an account of what the radical right seems to have in mind.

GOD. Many conservatives start with a view of God that makes conservative ideology seem both natural and good. God is the ultimate strict father—all good and all powerful, at the top of a natural hierarchy in which morality is linked with power. God wants good people to be in charge. Virtue is to be rewarded—with power. God therefore wants a hierarchical society in which there are moral authorities who should be obeyed in each domain: individual power, global power, financial power, social power.

God makes laws—commandments—defining right and wrong. One must have discipline to follow God's commandments. God is punitive: He punishes those who do not follow his commandments, and rewards those who do. Following God's laws takes discipline. Those who are disciplined enough to be moral are disciplined enough to become prosperous and powerful.

Christ, as savior, gives sinners a second chance—a chance to be born again and be obedient to God's commandments this time around.

THE MORAL ORDER. Traditional power relations are taken as defining a natural moral order: God above man, man above nature, adults above children, Western culture above non-Western culture, America above other nations. The moral order is all too often extended to men above women, whites above nonwhites, Christians above non-Christians, straights above gays.

MORALITY. Preserving and extending the conservative moral system (strict father morality) is the highest priority.

Morality comes in the form of rules, or commandments, made by a moral authority. To be moral is to be obedient to that authority. It requires internal discipline to control one's natural desires and instead follow a moral authority. What that authority is depends on your domain of interest: the individual, governing institutions—both public and private, Wall Street, conservative society.

Discipline is learned in childhood primarily through punishment for wrongdoing. Morality can be maintained only through a system of rewards and punishments.

ECONOMICS. Competition for scarce resources also imposes discipline, and hence serves morality. The discipline required to be moral is the same discipline required to win competitions and prosper.

The wealthy people tend to be the good people, a natural elite. The poor remain poor because they lack the discipline needed to prosper. The poor, therefore, deserve to be poor and serve the wealthy. The wealthy need and deserve poor people to serve them. The vast and increasing gap between rich and poor is thus seen to be both natural and good.

To the extent that markets are "free," they are a mechanism for the disciplined (stereotypically good) people to use their discipline to accumulate wealth. Free markets are moral: If everyone pursues his own profit, the profit of all will be maximized. Competition is good; it produces optimal use of resources and disciplined people, and hence serves morality. Regulation is bad; it gets in the way of the free pursuit of profit. Wealthy people serve society by investing and giving jobs to poorer people. Such a division of wealth ultimately serves the public good, which is to reward the disciplined and let the undisciplined be forced to learn discipline or struggle.

GOVERNMENT. Social programs are immoral. By giving people things they haven't earned, social programs remove the incentive to be disciplined, which is necessary for both morality and prosperity. Social programs should be eliminated. Anything that could be done by the private sphere should be. Government does have certain proper roles: to protect the lives and the private property of Americans, to make profit-seeking as easy as possible for worthy Americans (the disciplined ones), and to promote conservative morality (strict father morality), along with conservative social culture and religion.

EDUCATION. Since preserving and extending conservative morality is the highest goal, education should serve that goal. Schools should teach conservative values. Conservatives should gain control of school boards to guarantee this. Teachers should be strict, not nurturant, in the example they set for students and in the content they teach. Education should therefore promote discipline, and undisciplined students should face punishment. Unruly students should face physical punishment (for instance, paddling), and intellectually undisciplined students should not be coddled but should be shamed and punished by not being promoted. Uniform testing should test the level of discipline. There are right and wrong answers, and they should be tested for. Testing defines fairness: Those who pass are rewarded; those not disciplined enough to pass are punished.

Because immoral, undisciplined children can lead moral, disciplined children astray, parents should be able to choose to which schools they send their children. Government funding should be taken from public schools and given to parents in the form of vouchers. This will help wealthier (more disciplined and moral) citizens send their children to private or religious schools that teach conservative values and impose appropriate discipline. The vouchers given to poorer (less disciplined and less worthy) people will not be sufficient to allow them to get their children into the better private and religious schools. Schools will thus come to reflect the natural divisions of wealth in society. Of course, students who show exceptional discipline and talent should be given scholarships to the better schools. This will help maintain the social elite as a natural elite.

HEALTH CARE. It is the responsibility of parents to take care of their children. To the extent that they cannot, they are not living up to their individual responsibility. No one has the responsibility of doing other people's jobs for them. Thus prenatal care, postnatal care, health care for children, and care for the aged and infirm are matters of individual responsibility. They are not the responsibility of taxpayers.

SAME-SEX MARRIAGE AND ABORTION. Same-sex marriage does not fit the strict father model of the family; it goes squarely against it. A lesbian marriage has no father. A gay marriage has "fathers" who are taken to be less than real men. Since preserving and extending the strict father model is the highest moral value for conservatives, same-sex marriage constitutes an attack on the conservative value system as a whole, and on those whose very identity depends on their having strict father values.

Abortion works similarly. There are two stereotypical cases where women need abortions: unmarried teenagers who have been having "illicit" sex, and older women who want to delay child rearing to pursue a career. Both of these fly in the face of the strict father model. Pregnant teenagers have violated the commandments of the strict father. Career women challenge the power and authority of the strict father. Both should be punished by bearing the child; neither should be able to avoid the consequences of their actions, which would violate the strict father model's idea that morality depends on punishment. Since conservative values in general are versions of strict father values, abortion stands as a threat to conservative values and to one's identity as a conservative.

Conservatives who are "pro-life" are mostly, as we have seen, against prenatal care, postnatal care, and health care for children, all of which have major causal effects on the life of a child. Thus they are not really pro-life in any broad sense. Conservatives for the most part are using the idea of terminating a pregnancy as part of a cultural-war strategy to gain and maintain political power.

Both same-sex marriage and abortion are stand-ins for the general strict father values that define for millions of people their identities as conservatives. That is the reason why these are such hot-button issues for conservatives.

To understand this is not to ignore the real pain and difficulty involved in decisions made by individual women to terminate a pregnancy. For those truly concerned with the lives and health of children, the decision to end a pregnancy for whatever reason is always painful and anything but simple. It is this pain that conservatives are exploiting when they use ending pregnancy as a wedge issue in the cultural civil war they have been promoting.

There are also those who are genuinely pro-life, who believe that life begins with conception, that life is the ultimate value, and who therefore support prenatal care, postnatal care, health insurance for poor children, and early childhood education, and who oppose the death penalty, war, and so on. They also recognize that any woman choosing to end a pregnancy is making a painful decision, and empathize with such women and treat them without a negative judgment. These are pro-life progressives—often liberal Catholics. They are not conservatives who use the question of ending pregnancy as a political wedge to gain support for a broader moral and political agenda.

NATURE. God has given man dominion over nature. Nature is a resource for prosperity. It is there to be used for human profit.

CORPORATIONS. Corporations exist to provide people with goods and services, and to maximize profits for investors and upper management. They work most "efficiently" when they do maximize their profits. When corporations profit, society profits.

REGULATION. Government regulation stands in the way of free enterprise, and should be minimized.

RIGHTS. Rights must be consistent with morality. Strict father morality defines the limits of what is to count as a "right."

Thus there is no right to an abortion, no right to same-sex marriage, no right to health care (or any other government assistance), no right to know how the administration decides policy, no right to a living wage, and so on. However, there is a right to owning guns—especially conservatives owning guns—since guns provide a form of authority to those who possess them.

DEMOCRACY. A strict father democracy is an institutional democracy operating under strict father values. It counts as a democracy in that it has elections, a tripartite government, civilian control of the military, free markets, basic civil liberties, and widely accessible media. But strict father values are seen as central to democracy—to the empowerment of individuals to change their lives and their society by pursuing their individual interests.

FOREIGN POLICY. America is the world's moral authority. It is a superpower because it deserves to be. Its values—the right values—are defined by strict father morality. If there is to be a moral order in the world, American sovereignty, wealth, power, and hegemony must be maintained and American values—conservative family values, the free market, privatization, elimination of social programs, domination of man over nature, and so on—spread throughout the world.

THE CULTURE WAR. Strict father morality defines what a good society is. The very idea of a conservatively defined good society is threatened by liberal and progressive ideas and programs. That threat must be fought at all costs. The very fabric of society is at stake.

Those are the basics. Those are the ideas and values that the right wing wants to establish, nothing less than a radical revolution in how America and the rest of the world functions. That the vehemence of the culture war is provoked and maintained by conservatives is no accident. For strict father morality to gain and maintain political power, disunity is required. First, there is economic disunity, the two-tier economy with the "unworthy" poor remaining poor and serving the "deserving" rich. But to stay in power, conservatives need the support of the conservative poor. That is, they need a significant percentage of the poor and middle class to vote against their economic interests and for their individual, social, and religious interests. This means that what appears to be a division among conservatives on the basis of domains of interest actually constitutes strength for conservatism on the whole. Conservatism in all those domains of interest is required for conservatism to reign.

This has been achieved through the recognition that many working people and evangelical Protestants have a strict father morality in their families or religious lives. Conservative intellectuals have realized that these are the same values that drive political conservatism. They have also realized that people vote their values and their identities more than their economic self-interests. What they have done is to create, via framing and language, a link between strict father morality in the family and religion on the one hand and conservative politics and business on the other. This conceptual link must be so emotionally strong in those who are not wealthy that it can overcome economic self-interest.

Their method for achieving this has been the cultural civil war—a civil war carried out with everything short of live ammunition—pitting Americans with strict father morality (called conservatives) against Americans with nurturant parent morality (the hated liberals), who are portrayed as threatening the way of life and the cultural, religious, and personal identities of conservatives.

Conservative political and intellectual leaders faced a challenge in carrying out their goals. They represented an economic and political elite, but they were seeking the votes of middle- and lower-class working people. They needed, therefore, to identify conservative ideas as populist and liberal or progressive ideas as elitist—even though the reverse was true. They faced a massive framing problem, a problem that required a change in everyday language and thought. But strict father morality gave them an important advantage: It suggests that the wealthy have earned their wealth, that they are good people who deserve it—and that those who govern, both in the public and private sphere, should maintain the right moral order in society. It is a kind of conservative social contract.

Through the work of their think tank intellectuals, their language professionals, their writers and ad agencies, and their media specialists, conservatives have worked a revolution in thought and language over forty to fifty years. Through language they have branded liberals (whose policies are populist) as effete elitists, unpatriotic spendthrifts—using terms like limousine liberals, latte liberals, tax-and-spend liberals, Hollywood liberals, East Coast liberals, the liberal elite, wishy-washy liberals, and so on. At the same time they have branded conservatives (whose policies favor the economic elite) as populists—again through language, including body language. From Ronald Reagan's down-home folksiness to George W. Bush's John Wayne–style "Bubbaisms," the language, dialects, body language, and narrative forms have been those of rural populists. Their radio talk show hosts—warriors all—have adopted the style of hellfire preachers. But the message is the same: The hated liberals are threatening American culture and values, and have to be fought vigorously and continuously on every front. It is a threat to the very security of the nation, as well as morality, religion, the family, and everything real Americans hold dear. Their positions on wedge issues—guns, babies, taxes, same-sex marriage, the flag, school prayer—reveal the "treachery" of liberals. The wedge issues are not important in themselves, but are vital in what they represent: a strict father attitude to the world.

Without the mutually supportive relationships among the domains of interest—individual, governing, business, and social—conservatism as an overarching moral system cannot flourish. What appears to liberals to be fragmentation and disunity could be a mutually reinforcing structure that is powerful and threatening to progressive values and American democracy.

The prevailing wisdom of progressives, on the other hand, is that ideological divisions are tearing conservatism apart. But we had better consider both possibilities.

14

What Unites Progressives

To approach what unites progressives, we must first ask what divides them. Here are some of the common parameters that divide progressives from one another:

- **Local interests.** For example, you might come from a farming community, or a high-tech region, or a city with a military base, or a home base with a large racial or ethnic minority population—and you place the concerns of that region high on your priorities.
- **Idealism versus pragmatism.** As a pragmatist, you are willing to compromise and get the best deal you can; as an idealist, you may be unwilling to compromise. Idealists tend to accuse pragmatists of not having ideals (when they do, but can't realize them); pragmatists criticize idealists, saying "the perfect is the enemy of the good."
- **Biconceptualism.** If you are mostly progressive but have some conservative views, total progressives will accuse you of being a conservative; biconceptuals tend to accuse total progressives of being dogmatic or extremists.
- **Radical change versus gradual change.** Radicals accuse gradualists of not being truly progressive; gradualists accuse radicals of being impractical and hurting their own cause by not using slippery slope tactics.
- **Militant versus moderate advocacy.** Militants are loud, aggressive, and punitive, and sometimes use strict father means to nurturant ends, and see moderates as being cowards or insufficiently caring; moderate advocates think that militancy offends people and causes a reaction against their cause.
- **Types of thought processes.** Progressive values can be weighted toward different areas of concern: socioeconomic, identity politics, environmentalist, civil libertarian, spiritual,

and antiauthoritarian (see *Moral Politics* for details). Each thought process has consequences in choosing what causes to pursue, how to rank priorities, how to use political capital, where and how to raise money and what to spend it on, who your friends and acquaintances are, what to read, who to pay attention to, and so on.

A great many progressives have been critical of President Obama. If you were to make a list of the criticisms, they would mostly be defined by one or more of these parameters: too pragmatic, not a real progressive, going too slow, too timid or cowardly, not militant enough, not doing enough for *my* major interest.

When you consider that each progressive has some distinct combination of these parameters, the number of types of progressives becomes astronomical. When Will Rogers said, "I am not a member of any organized political party. I am a Democrat," this is the kind of thing he meant.

This makes it all the more important to understand what unites progressives and how to openly discuss that unity—despite the differences defined by these parameters.

Programs are a major problem for attempts at unity. As soon as a policy is made specific, the differences must be addressed. Progressives tend to talk about policies and programs. But policy details are not what most Americans want to know about. Most Americans want to know what you stand for, whether your values are their values, what your principles are, what direction you want to take the country in. In public discourse, values trump policies, principles trump policies, policy directions trump specific programs. I believe that values, principles, and policy directions are exactly the things that can unite progressives, if they are crafted properly. The reason that they can unite us is that they stand conceptually above all the things that divide us.

Ideas That Make Us Progressives

What follows is a detailed explication of each of those unifying ideas:

- First, *values* coming out of a basic progressive vision
- Second, *principles* that realize progressive values
- Third, *policy directions* that fit the values and principles

The Basic Progressive Vision

The basic progressive vision is of community—of America as family, a caring, responsible family. We envision an America where people care about each other, not just themselves, and act responsibly both for themselves and their fellow citizens with strength and effectiveness.

Democracy means acting on that care and responsibility through the government to provide public resources for all—from the needy to the average citizen to those running businesses, great or small. In short, the private depends on the public. And if you used those public resources to become wealthy on the basis of taxes paid by others for the resources you used, then fairness requires that you pay a higher share of your wealth in taxes so that others may benefit as well.

We are all in the same boat—that's what democracy means. Red states and blue states, progressives and conservatives, Republicans and Democrats. United, as we were for a brief moment just after September 11, not divided by a despicable culture war.

So far as I know, every progressive shares those values and that view of democracy.

The Logic of Progressive Values

The progressive core values are family values—those of the responsible, caring family. You could characterize them as **caring and responsibility, carried out with strength of commitment and effort.** These core values imply the full range of progressive values, listed below, together with the logic that links them to the core values.

- **Protection, fulfillment in life, fairness.** When you care about someone, you want them to *be protected from harm*, you want their *dreams to come true*, and you want them to be *treated fairly*.
- **Freedom, opportunity, prosperity.** There is no *fulfillment* without *freedom*, no *freedom* without *opportunity*, and no *opportunity* without *prosperity*.

- **Community, service, cooperation.** Children are shaped by their *communities*. Responsibility requires *serving* and helping to shape your community. That requires *cooperation*.
- **Trust, honesty, open communication.** There is no *cooperation* without *trust*, no *trust* without *honesty*, and no *cooperation* without *open communication*.

Just as these values follow from caring and responsibility, so every other progressive value follows from these. Equality follows from fairness, empathy is part of caring, diversity comes from empathy and equality.

Progressive Principles

Progressives not only share these values but also share political principles that arise from these values.

EQUITY. What citizens and the nation owe each other. If you work hard, play by the rules, and serve your family, community, and nation, then the nation should provide a decent standard of living as well as freedom, security, and opportunity.

EQUALITY. Do everything possible to guarantee political and social equality and avoid imbalances of political power.

DEMOCRACY. Maximize citizen participation; minimize concentrations of political, corporate, and media power. Maximize journalistic standards. Establish publicly financed elections. Invest in public education. Bring corporations under stakeholder control, not just stockholder control.

GOVERNMENT FOR A BETTER FUTURE. Government does what America's future requires and what the private sector cannot do—or is not doing—effectively, ethically, or at all. It is the job of government to promote and, if possible, provide sufficient protection, greater democracy, more freedom, a better environment, broader prosperity, better health, greater fulfillment in life, less violence, and the building and maintaining of public infrastructure.

ETHICAL BUSINESS. Our values apply to business. In the course of making money by providing products and services, businesses should not adversely affect the public good, as defined by the above values. They should refuse to impose wage slavery and corporate servitude and so should work with unions, not against them. They should pay the true costs of doing business—not externalize, or offload, those costs onto the public (for instance, they should clean up their pollution). They should make sure their products do no harm to the public. And rather than treat their employees as mere "resources," they should see them as community members and assets to the business.

VALUES-BASED FOREIGN POLICY. The same values governing domestic policy should apply to foreign policy whenever possible.

Here are a few examples where progressive domestic policy translates into foreign policy:

- Protection translates into an effective military for defense and peacekeeping.
- Building and maintaining a strong community translates into building and maintaining strong alliances and engaging in effective diplomacy.
- Caring and responsibility translate into caring about and acting responsibly for the world's people; helping to deal with problems of health, hunger, poverty, and environmental degradation; population control (and the best method, women's education); and rights for women, children, prisoners, refugees, and ethnic minorities.

All of these are concerns of a values-based foreign policy.

Policy Directions

Given progressive values and principles, progressives can agree on basic policy directions, if not on the details. Policy directions are at a higher level than specific policies. Progressives usually divide on specific policy details while agreeing on directions. Here are some of the many policy directions they agree on.

THE ECONOMY. Investing in an economy centered on innovation that creates millions of good-paying jobs and provides every

American a fair opportunity to prosper. The economy should be sustainable and not contribute to climate change, environmental degradation, and so on.

SECURITY. Through military strength, strong diplomatic alliances, and wise foreign and domestic policy, every American should be safeguarded at home, and America's role in the world should be strengthened by helping people around the world live better lives.

HEALTH. Every American should have access to a state-of-the-art, affordable health care system.

EDUCATION. A vibrant, well-funded, and expanding public education system, with the highest standards for every child and school, where teachers nurture children's minds and often the children themselves, and where children are taught the truth about their nation—its wonders and its blemishes.

EARLY CHILDHOOD. Every child's brain is shaped crucially by early experiences. We support high-quality early childhood education.

ENVIRONMENT. A clean, healthy, and safe environment for ourselves and our children: water you can drink, air you can breathe, and food that is healthy and safe. Polluters pay for the damage they cause.

NATURE. The natural wonders of our country are to be preserved for future generations, and enhanced where they have been degraded.

ENERGY. We need to make a major investment in renewable energy, for the sake of millions of jobs that pay well, improvements in public health, preservation of our environment, and the effort to halt global warming.

OPENNESS. An open, efficient, and fair government that tells the truth to our citizens and earns the trust of every American.

EQUAL RIGHTS. We support equal rights in every area involving race, ethnicity, gender, and sexual orientation.

PROTECTIONS. We support keeping and extending protections for consumers, workers, retirees, and investors.

A **stronger America** is not just about defense, but about every dimension of strength: our effectiveness in the world, our economy, our educational system, our health care system, our families, our communities, our environment, and so forth.

Broad prosperity is the effect that markets are supposed to bring about. But all markets are constructed for someone's benefit; no markets are completely free. Markets should be constructed for the broadest possible *prosperity*, as opposed to the exponential accumulation of wealth by the wealthy coupled with the corresponding loss of wealth by most citizens—and with it the loss of freedom and fulfillment in life.

Americans want and deserve a **better future**—economically, educationally, environmentally, and in all other areas of life—for themselves and their children. Lowering taxes, primarily for the super-rich elite, has had the effect of defunding programs that would make a better future possible in all these areas. The proper goal is a better future for all Americans. This includes bringing global warming under control.

Smaller government is, in conservative propaganda, supposed to eliminate waste. It is really about eliminating social programs. **Effective government** is what we need our government to accomplish to create a better future.

We should be governed not by corporations, but by a government of, by, and for the people.

Conservative family values are those of a strict father family—authoritarian, hierarchical, every man for himself, based around discipline and punishment. Progressives live by the best values of both families and communities: **mutual responsibility**, which is authoritative, equal, and based around caring, responsibility (both individual and social), and commitment.

The remarkable thing is just how much progressives do agree on. These are just the things that voters tend to care about most: our values, our principles, and the direction in which we want to take the nation.

I believe that progressive values *are* traditional American values, that progressive principles are fundamental American principles, and that progressive policy directions point the way to where most Americans really want our country to go. The job of unifying progressives is really the job of bringing our country together around its finest traditional values.

But having those shared values, largely unconscious and unspoken, is not good enough. They have to be out in the open, named, said, discussed, publicized, and made part of everyday public discourse. If they go unspoken, while conservative values dominate public discourse, then those values can be lost—swept out of our brains by the conservative communication juggernaut.

Don't just read about these values here and nod. Get out and say them out loud. Discuss them wherever you can. Volunteer for campaigns that give you a chance to discuss these values loud and clear and out in public.

15

Frequently Asked Questions

Any brief discussion of framing and moral politics will leave many questions unanswered. Here are some of the most common questions I've been asked.

There is an asymmetry between *strict father* and *nurturant parent*. Why is the first masculine and the second gender-neutral?

In the strict father model, the masculine and feminine roles are very different, and the father is the central figure. The strict father is the moral authority of the family, the person in charge of the family, while mothers are seen as being "mommies"—they may be loving, but they are unable to protect and support the family and aren't strict enough to punish their children when they do wrong. Think of the expression "Wait till Daddy gets home," which refers to a strict daddy.

In this strict father model, "mommies" are supposed to uphold the authority of the strict father, but they are not able to do the job themselves. In the nurturant parent model, there just isn't a gender distinction of this sort. Both parents are there to nurture their children and to raise them to be nurturers. That doesn't mean there won't be gender-based divisions of labor around the house in real life, but they are not within the nurturant parent model.

These models are, of course, stereotypes—idealized, incomplete, oversimplified mental models. Mental models of this sort necessarily differ from real-world cases: strict mothers, single-parent households, gay parents, and so on.

Conservative commentators like David Brooks have referred to the Republicans as the "daddy party" and the Democrats as the "mommy party." Would you agree?

Brooks and others have acknowledged the Nation as Family metaphor, and have acknowledged that the strict father model is behind conservative Republican politics. However, their characterization of a "mommy party" is based on "mommy" in their own conservative, strict father model. What they mean by "mommy party" is that although Democrats may care and be loving people, they just aren't tough and realistic enough to do the job.

This is, of course, completely inaccurate from the Democrats' own progressive perspective. In a nurturant family, both parents are not just caring but also responsible and strong enough to carry out those responsibilities. This is far from *mommy* in the way the conservatives scornfully use the term. Democrats have been able to successfully provide both protection for and prosperity to the nation.

Conservatives seem not to understand what nurturant morality is about, both in the family and in the nation. They find any view that is not strict to be "permissive." Nurturant parenting is, of course, anything but permissive, with its stress on teaching children to be responsible for themselves and empathetic and responsible toward others, and raising them to be strong and well-educated enough to carry out their responsibilities. The conservatives parody liberals as permissive, as supporting a feel-good morality—doing whatever feels good. The conservatives just don't get it. They seem ignorant of the vast difference between responsibility and permissiveness.

How old are the ideas of strictness and nurturance?

They seem to go back very, very far in history. We know, for example, that in England before the British came over to colonize America there were religious groups like the Quakers, who had a nurturant view of God, and groups like the Puritans, who had a strict father view of God. The New England colonies were mainly Puritan, though John Winthrop had a nurturant view of the colony he was establishing, and the nurturant view of God has existed side by side with the strict one in this country ever since. In the nineteenth century, Horace Bushnell wrote about "Christian nurture." From the period of the abolitionists through the 1920s there was a lively discussion of the nurturant view of God. Moreover, students of religion have shown that there are strict and nurturant views of religion that go back as far as biblical and prebiblical times. These distinctions have been there for a very, very long time.

Does the strict father model imply that conservatives don't love their kids, and does the nurturant parent model imply that progressives don't believe in discipline?

Not at all. In the strict father model, physically disciplining a child who has done wrong, by inflicting sufficient pain, is a form of love—"tough love." Given the duty to impose "loving discipline," lots of hugging and other loving behavior are permissible, and often recommended, afterward. It's just a matter of first things first.

In the nurturant parent model, discipline arises not through painful physical punishment but through the promotion of responsible behavior via empathetic connection, the example of responsible behavior set by the parents, the open discussion of what the parents expect (and why), and, in the case of noncooperation, the removal of some of the good things that go with cooperation. A child raised through nurturance is a child who has achieved positive internal discipline without painful physical punishment. It is achieved through praise for cooperation, understanding the privileges that go with cooperation, clear guidelines, open discussion, and the example of parents who live by their nurturant values.

What are the complexities of the models?

The models (discussed in detail in chapter 17 of *Moral Politics*), have built-in complexities.

First, just about everybody in American culture has both models, either actively or passively. For example, to understand a John Wayne movie, you must have a strict father model in your brain, at least passively. You may not live by the model, but you can use it to understand the strict father narratives that permeate our culture. Nurturant narratives permeate our culture as well.

Second, many people use both models, but in different parts of their lives. For example, a lawyer might be strict in the courtroom but nurturant at home.

Third, you may have been brought up badly with one model, and may have rejected it. Many liberals had miserable strict father upbringings.

Fourth, there are three natural dimensions of variation for applying a given model: an ideological/pragmatic dimension, a radical/moderate dimension, and a means/ends dimension.

Both a progressive and a conservative can be unyielding ideologues, or they may be pragmatic—willing to compromise on a proposal either for reasons of real-world workability or political viability.

In addition, both progressives and conservatives can vary on the two radical/moderate scales: the amount of change and the speed of change. Thus radical conservative ideologues are unwilling to compromise, and insist on the most rapid and complete change possible.

Incidentally, the word *conservative* is not necessarily about conserving anything. It is about strict father morality. There is no contradiction in talking about "radical conservatives." Indeed, Robert Reich, in his book *Reason*, uses the term *radcon* to talk about radical conservatives. From this perspective a "moderate" can be either a progressive or a conservative who is pragmatic or wants slow change, a bit at a time. It is sometimes said that there is a third moderate model, very different from the other two, but I have not yet seen such a model proposed explicitly.

Another common variation occurs in distinguishing ends and means. There are people with progressive politics (nurturant ends) who have strict father means. These are the militant progressives. The most extreme case is the authoritarian antiauthoritarians: those with antiauthoritarian progressive ends but authoritarian strict father organizations.

Last, there are the types—the special cases—of progressives and conservatives that we discussed in chapter 1: the socioeconomic, identity politics, environmentalist, civil libertarian, antiauthoritarian, and spiritual progressives; and the financial, social, libertarian, neocon (see chapter 13), and religious conservatives. They are all instances of the nurturant and strict models, but each restricts the form of reasoning used.

The notion of reframing sounds manipulative. How is framing different from spin or propaganda?

Framing is normal. Every sentence we say is framed in some way. When we say what we believe, we are using frames that we think are relatively accurate. When a conservative uses the "tax relief" frame, chances are that he or she really believes that taxation is an affliction. However, frames can also be used manipulatively. The use, for example, of "Clear Skies Act" to name an act that increases air pollution is a manipulative frame. And it's used to cover up a weakness that conservatives have, namely that the public doesn't like legislation that increases air pollution, and so they give it a name that conveys the opposite frame. That's pure manipulation.

Spin is the manipulative use of a frame. Spin is used when something embarrassing has happened or has been said, and it's an attempt to put an innocent frame on it—that is, to make the embarrassing occurrence sound normal or good.

Propaganda is another manipulative use of framing. Propaganda is an attempt to get the public to adopt a frame that is not true and is known not to be true, for the purpose of gaining or maintaining political control.

The reframing I am suggesting is neither spin nor propaganda. Progressives need to learn to communicate using frames that they really believe, frames that express what their moral views really are. I strongly recommend against any deceptive framing. I think it is not just morally reprehensible, but also impractical, because deceptive framing usually backfires sooner or later.

Why don't progressives take advantage of wedge issues?

Conservatives have been thinking about the strategic use of ideas; progressives haven't, but we could. We could perfectly well use wedge issues. They're all around us. Take something like clean air and clean water. Conservatives want clean air and clean water. That can be made into a wedge issue.

Imagine a campaign for poison-free communities, starting with mercury as the poison of choice, then going on to other kinds of poison in our air and in our water, around us in various forms. That could be made into an effective wedge issue, splitting the conservatives who care about their own and their children's health from those who are simply against government regulation. The very issue would create a frame in which regulation favors health, and being against regulation endangers health.

This is also a slippery slope issue. Once you get people looking at how and where mercury enters the environment—for example, from the processing of coal and many other kinds of chemicals—and you get people thinking about cleaning up mercury, and about mercury poisoning, and how it works in the environment, you can move to the next poison in the environment, and the poison after that, and the poison after that.

This is an issue that is not just about mercury or about poisons in the environment, but about nurturant morality in general. Wedge issues are stand-ins for the whole of a moral system. Abortion is an issue that serves as a stand-in for the control of women's lives and for a moral hierarchy that conservatives want to impose. Abortion, as we have seen, is a stand-in for strict father morality in general. Similarly, there are all sorts of wedge issues that can be stand-ins for progressive morality in general.

Is religion inherently conservative? Are progressive ideals inconsistent with *religious beliefs*?

Conservatives would have us believe that religions are conservative, but they're not. Millions of Christians in this country are liberal Christians. Most Jews are liberal Jews. And I suspect that most Muslims in America are progressive, liberal Muslims, not radically conservative Muslims. However, the progressive religious community in this country is not well organized, while the conservative religious community is extremely well organized. One of the problems is that the progressive religious community, particularly progressive Christianity, doesn't really know how to express its own theology in a way that makes its politics clear, whereas conservative Christians do know the direct link between their theology and their politics. Conservative Christianity is a strict father religion. Here's how the strict father view of the world is mapped onto conservative Christianity.

First, God is understood as punitive—that is, if you sin you are going to go to hell, and if you don't sin you are going to be rewarded and go to heaven. But since people tend to sin at one point or another in their lives, how is it possible for them to ever get to heaven? The answer in conservative Christianity is Christ. What Jesus does is offer conservative Christians a chance to get to heaven. The idea is this: Christ suffered on the cross so much that he built up moral credit sufficient for all people, forever. He then offered a chance to get to heaven—that is, redemption—on the following terms, strict father terms: If you accept Jesus as your savior, that is, as your moral authority, and agree to follow the moral authority of your minister and your church, then you can get to heaven. But that is going to require discipline. You need to be disciplined enough to follow the rules, and if you don't, then you are going to go to hell. So Jesus, with his moral credit that he gained from suffering, can pay off your debts—that is, your sins—and allow you to get into heaven, but only if you toc the line.

Liberal Christianity is very, very different. Liberal Christianity sees God as essentially beneficent, as wanting to help people. The central idea in liberal Christianity is grace, where grace is understood as a kind of metaphorical nurturance. In liberal Christianity, you can't earn grace—you are given grace unconditionally by God. But you have to accept grace, you have to be near God to get his grace, you can be filled with grace, you can be healed by grace, and you are made into a moral person through God's grace.

In other words, grace is metaphorical nurturance. That is, just as nurturance feeds you, heals you, takes care of you, just as a nurturant parent teaches you to be nurturant and allows you to be a moral being, just as you can't get nurturance unless you are close to your parents, just as you must accept nurturance in order to get it, so all of these things about nurturance are true of grace in liberal Christianity. Nurturance comes with unconditional love—in the case of grace, the unconditional love of God. What makes a religion nurturant is that it metaphorically views God as a nurturant parent. In a nurturant form of religion, your spiritual experience has to do with your connection to other people and the world, and your spiritual practice has to do with your service to other people and to your community. This is why nurturant Christians are progressives: because they have a nurturant morality, just as progressives have.

But at present nurturant Christians, Jews, Muslims, Buddhists, and others in this country are not organized. They are not seen as a single movement, a progressive religious movement. Worse, secular progressives do not see those with a nurturant form of religion as natural members of the same political movement. Not only do spiritual progressives need to unite with each other, they need to unite with secular progressives, who share the same moral system and political objectives.

What is a strategic initiative, and how is it different from regular policy making?

There are two kinds of strategic initiatives: The first is what I call a slippery slope initiative. The idea of a slippery slope initiative is to take a first step that seems fairly straightforward, but gets into the public eye an additional frame that you want to be there. The idea is that once the first step is taken, then it is easier and often inevitable to take the next step and the next step and the next step.

The conservative Supreme Court works by slippery slope decisions, one step at a time. Consider the following progression. First, the court allowed corporations to contribute to ballot initiatives as a limited form of the First Amendment right of free speech. Then, their Citizens United decision gave corporations the ability to contribute as much as they want in elections, as a form of free speech. Then their Hobby Lobby decision extended the First Amendment freedom of religion to corporations so that they do not have to provide contraception to women employees as specified by the Affordable Care Act, opening the door to a wider use of freedom of religion by corporations to avoid various fair treatment laws.

Let's take another example. It used to be the case that conservatives tried to cut social programs one by one, and then they figured out how they could cut them all at once: through tax cuts. Cutting taxes is a strategic initiative, not of the slippery slope variety but of a deeper variety, one that has wide effects across many, many areas. If you cut taxes and create a large deficit, then when any social program comes up—it could be health care for poor children, or services for paraplegics, or whatever—there won't be enough money for it. So you end up cutting social programs across the board in health, in education, in the enforcement of environmental regulations, and so on. At the same time you reward those who you see as the good people, namely the wealthy people—those who were disciplined enough to become wealthy.

There are other kinds of strategic initiatives as well. Take the example of same-sex marriage. Same-sex marriage contradicts large parts of the strict father model. If it's a lesbian marriage, there's no father at all, and in a gay marriage, where there are two fathers, neither of them fits the traditional view of the male strict father. Opposing same-sex marriage is thus reinforcing and extending strict father morality itself, which is the highest calling of the conservative moral system. Same-sex marriage is therefore a stand-in; it evokes the larger issue, namely what moral system is to govern our country.

The same is true of the issue of abortion. Allowing women to decide for themselves on whether to end a pregnancy flies in the face of the whole idea of a strict father family model. In the strict father model, it is the father who decides whether his wife or daughter should have an abortion. It is the father who controls his daughter's sexuality; when the daughter takes a lover, then the father loses control. If the father is to maintain control over his family, then the women in the family cannot freely control their own sexual behavior and their own ability to reproduce. Abortion is therefore not inherently a political issue, but only a political issue when it comes to whether strict father morality is to reign in American life. Abortion is a stand-in for the larger issue: Is strict father morality going to rule America?

So all I have to do to reframe my issue is think up some sound bite–worthy terms and use them in place of the conservative terms?

No! Reframing is not just about words and language. Reframing is about ideas. The ideas have to be in place in people's brains before the sound bite can make any sense. For example, take the idea of "the commons"—that is, our common inheritance, like the atmosphere or the electromagnetic spectrum (bandwidths). These are the common inheritances of all humanity, and most people who discuss them in this way refer to them as "the commons." Yet the idea of a common inheritance and of using it for the public good is not yet part of the frame structure that most people use every day. For this reason you can't just make up a sound bite about the commons and have most people understand it and agree with it.

If Republicans have such a huge infrastructure, how do we catch up?

Progressives know that they have to make investments in media. What they tend not to know is that they have to make investments in framing and in language. The big advantage we have is this: Whereas it took more than thirty years, billions of dollars, and forty-three institutes for conservatives to reframe public debate so the debate occurs on their turf, we have the advantage of having science on our side. Through cognitive science and through linguistics, we know how they did it. And we know how we can do the equivalent for progressives in a much shorter time and with many fewer resources. We also know how they've done their linguistic training, and we know how to do it ourselves.

Unfortunately, many progressives think this can be done through ad agencies and through pollsters. That's a mistake. You really do need linguists and cognitive scientists, platforms for in-depth and sustained discussion, and well-honed plans for keeping meaningful dialogue consistently before policy makers and the public.

What was the difference between the Rockridge Institute and other progressive think tanks? Are there any other think tanks that are dedicated to research on framing?

Rockridge was entirely dedicated to reframing the public debate, both from a policy perspective and from a linguistic perspective. Other progressive think tanks have other primary functions: responding to the initiatives of the right, answering conservative charges, telling the truth when there are conservative lies, and constructing specific policies that progressives can use. All of these are important functions, but they do not replace the framing function, a function that is absolutely necessary.

To my knowledge, there is now only one think tank devoted to the overall framing of issues from both a policy perspective and a communicative perspective—the Forward Institute in Wisconsin. The Forward Institute is dedicated to empowering the progressives of Wisconsin to frame state issues from a progressive viewpoint. They have studied the framing of Wisconsin issues and have trained trainers to work with a full range of progressives—from political leaders at all levels, to union leaders, to teachers, to Native Americans, to environmentalists, to citizen-volunteers all willing to speak around the state using progressive frames. The institute has just started. Only time will tell if they will get the funding they need to succeed.

Isn't *tax relief* the natural way to talk about taxes? I'm a progressive, but I have to admit, they do seem burdensome sometimes.

Homework in school is burdensome too, but you have to do it if you're going to learn anything. Exercise is burdensome, but you have to do it if you're going to be in good physical shape. Taxes are necessary if we are going to make wise investments in our national infrastructure that will pay off for all of us years and years in the future. That includes investments in things like education and health care for those who can't afford them. Education and health care are investments in people. They are wise investments because they give us an educated citizenry, an educated workforce, and a healthy and efficient workforce. Those are the practical reasons for taxes. Other reasons for taxes are public services—like police and fire, disaster relief, and so on.

There are moral reasons for taxes as well. Education and health are important factors in fulfillment in life, and this country is about fulfillment in life. There is a reason why the Declaration of Independence talks about the pursuit of happiness and links it to liberty. The reason is that they go together. Without liberty, there can be no fulfillment in life. Thus there are practical reasons why it makes sense to understand taxation as investment, and there are moral reasons to understand taxation as paying your dues in a country where you can pursue happiness because there is liberty and freedom.

How do you respond or reply directly to a Republican strategic initiative?

You can't, and that's why they're clever. Tax cuts are not about tax cuts. That's why you can't reply directly to tax cuts so easily. They are about getting rid of all social programs and regulations of business. Vouchers and school testing are not ultimately about vouchers and school testing; they are about conservative control of the content of education and the elimination of public resources. To respond you have to put the individual issue into a much larger framework that fits your understanding of the situation. Tort reform is not about tort reform; it is about allowing corporations to act without restraints, and about taking funding away from the Democratic Party, since trial lawyers are a major source of Democratic funding.

Instead of trying to reply to strategic initiatives, you need to reframe the larger issues at stake from your point of view. You can discuss the strategic initiative, or at least some parts of it, from your framework. Take tort reform. Trial lawyers are really public protection attorneys, and tort law is law that allows for public protection—it's public protection law. When tort law tries to cap claims and settlements, its effect is to take claims out of the hands of juries—that is, to close the courtroom door, to create closed courts instead of open courts. In open courts, where there are juries, the jury can decide whether a given claim is a matter of public protection. Large settlements often have to do with issues of public protection—that is, they go beyond the case at hand. And open courts are the last defense that the public has against unscrupulous or negligent corporations or professionals. When conservatives talk about the lawsuits, you don't just say, "No, no, the lawsuits weren't frivolous," you talk instead about public protection, about open courts, about the right to have juries decide, and about the last line of defense against unscrupulous or negligent corporations.

If facts that don't fit frames are rejected, does that mean we should stop using facts in our arguments?

Obviously not. Facts are all-important. They are crucial. But they must be framed appropriately if they are to be an effective part of public discourse. We have to know what a fact has to do with moral principles and political principles. We have to frame those facts as effectively and as honestly as we can. And honest framing of the facts will entail other frames that can be checked with other facts.

How do progressive values differ from traditional American values?

They don't differ. Progressive values are traditional American values, all the values we are proud of.

We are proud of the victories for equality and against hierarchy: the emancipation of the slaves, women's suffrage, the union movement, the integration of the armed forces, the civil rights movement, the woman's movement, the environmental movement, and the gay rights movement.

We are proud of FDR's conception of government "for the people" and his rally for hope against fear.

We are proud of the Marshall Plan, which helped to erase the notion of "enemies."

We are proud of John Kennedy's call to public service, of Martin Luther King's insistence on nonviolence in the face of brutality, of Cesar Chavez's ability to bring pride and organization to the worst-treated of workers.

Progressive thought is as American as apple pie. Progressives want political equality, good public schools, healthy children, care for the aged, police protection, family farms, air you can breathe, water you can drink, fish in our streams, forests you can hike in, songbirds and frogs, livable cities, ethical businesses, journalists who tell the truth, music and dance, poetry and art, and jobs that pay a living wage to everyone who works.

Progressive activists—for living wages, women's rights, human rights, the environment, health, voter registration, and so on—are American patriots, working with unselfish dedication toward making a better world, a world that fits fundamental American values.

16

How to Respond to Conservatives

The earlier chapters are meant to explain what framing is and how it works through language and communication systems, what conservative and progressive worldviews are, what biconceptualism is, and what the deep issues are in framing. But sooner or later, you are on the front line called the dinner table. As my students regularly ask, "Thanksgiving is coming and I'm going to be eating dinner with my conservative relatives, and I am going to get in a row over politics with my grandfather or my aunt. It's always painful. What can I do?"

The following is a letter I received in 2004 while writing the version of this chapter in the first edition. It arrived several days after I had appeared on a TV show, *NOW with Bill Moyers*.

> I listened to Dr. Lakoff last Friday night on *NOW* with great interest. I love the use of words and have been consistently puzzled at how the far right has co-opted so many definitions.
>
> So I tried an experiment I wanted to tell you about. I took several examples from the interview; particularly trial vs. public protection lawyer and gay marriage and used those examples all week on AOL's political chat room. Every time someone would scream about [John] Edwards's being a trial lawyer, I'd respond with public protection lawyer and how they are the last defense against negligent corporations and [are] professional, and that the opposite of a public protection lawyer is a corporate lawyer who typically makes $400–500/per hr., and we pay that in higher prices for goods and services.
>
> Every time someone started screaming about "gay marriage" I'd ask if they want the federal government to tell them who they could marry. I'd go on to explain when

challenged that once government has crossed the huge barrier into telling one group of people who they could not marry, it is only a small step to telling other groups, and a smaller yet step to telling people who they had to marry.

I also asked for definitions. Every time someone would holler "dirty liberal," I'd request their definition of "liberal."

The last was my own hot button. Every time someone would scream "abortion," "baby-killer," etc., I'd suggest that if they are anti-abortion, then by all means, they should not have one.

I've got to tell you, the results were startling to me. I had some other people (completely unknown to me) join me and take up the same tacks. By last night, the chat room was civil. An amazing (to me) number of posters turned off their capitalization and we were actually having conversations.

I'm going to keep this up, but I really wanted you to know that I heard Dr. Lakoff, appreciate his work, and am trying to put it into practice. And it's really really fun.

Thanks,
Penney Kolb

This book is written for people like Penney Kolb. Progressives are constantly put in positions where they are expected to respond to conservative arguments. It may be over Thanksgiving dinner, around the water cooler, or in front of an audience. But because conservatives have commandeered so much of the language, progressives are often put on the defensive with little or nothing to say in response.

But sooner or later, you are in Penney's position. What do you do? Penney's instincts are impeccable, and provide us with guidelines.

- Progressive values are the best of traditional American values. Stand up for your values with dignity and strength. You are a true patriot because of your values.

- Remember that right-wing ideologues have convinced half of the country that the strict father family model, which is bad enough for raising children, should govern our national morality and politics. This is the model that the best in American values has defeated over and over again in the course of our history—from the emancipation of the slaves to women's suffrage, Social Security and Medicare, the civil rights and voting rights acts, *Brown v. the Board of Education*, and *Roe v. Wade*. Each time we have unified our country more behind our finest traditional values.

- Remember that most people have both strict and nurturant models, either actively or passively, perhaps active in different parts of their lives. Your job is to activate for politics the nurturant, progressive values already there (perhaps only passively) in whoever you're talking to.

What do I tell my students when they ask what to say at Thanksgiving dinner? My advice: Ask your aunt or grandfather what they are most proud of that helped other people. Those of my students who have done this report that, to their surprise, their grandfather or other relative did a number of good things to help others and show some important social concerns. My next bit of advice: Keep talking about those things. The more you keep talking about *their* empathy and responsibility toward others, the closer you can get to them. Don't try to convert them. Just try to open up and maintain a positive relationship. If you show respect and affection for your relatives, you may get some back.

- Be sure to show respect to the conservatives you are responding to. No one will listen to you if you don't accord them respect. Listen to them. You may disagree strongly with everything that is being said, but you should know what is being said. Be sincere. Avoid cheap shots. What if they don't show you respect? Two wrongs don't make a right. Turn the other cheek and show respect anyway. That takes character and dignity. Show character and dignity.

- Avoid a shouting match. Remember that the radical right requires a culture war, and shouting is the discourse form of that culture war. Civil discourse is the discourse form of nurturant morality. You win a victory when the discourse turns civil. They win when they get you to shout.

- What if you have moral outrage? You should have moral outrage. But you can display it with controlled passion. If you lose control, they win.

- Distinguish between ordinary conservatives and nasty ideologues. Most conservatives are personally nice people, and you want to bring out their niceness and their sense of neighborliness and hospitality.

- Be calm. Calmness is a sign that you know what you are talking about.

- Be good-humored. A good-natured sense of humor shows you are comfortable with yourself.

- Hold your ground. Always be on the offense. Never go on defense. Your voice should be steady. Never whine or complain. Your body and voice should show optimism. Never act like a victim. Never plead. You should convey passionate conviction without losing control. Avoid the language of weakness—for example, rising intonations on statements.

- Conservatives have parodied liberals as weak, angry (hence not in control of their emotions), weak-minded, softhearted, unpatriotic, uninformed, and elitist. Don't give them any opportunities to stereotype you in any of these ways. Expect these stereotypes, and deal with them when they come up.

- By the way you conduct yourself, show strength, calmness, and control; an ability to reason; a sense of realism; love of country; a command of the basic facts; and a sense of being an equal, not a superior. At the very least you want your audience to think of you with respect, as someone they may disagree with but who they have to take seriously. In many situations this is the best you can hope for. You have to recognize those situations and realize that a draw with dignity is a victory in the game of being taken seriously.

- Many conversations are ongoing. In an ongoing conversation, your job is to establish a position of respect and dignity, and then keep it.
- Don't expect to convert staunch conservatives.
- You can make considerable progress with biconceptuals, those who use both models but in different parts of their life. They are your best audience. With biconceptuals your goal is to find out, if you can by probing, just which parts of their life they are nurturant about. For example, ask who they care about the most, what responsibilities they feel they have to those they care about, and how they carry out those responsibilities. This should activate their nurturant models as much as possible. Then, while the nurturant model is active for them, try linking it to politics. For example, if they are nurturant at home but strict in business, talk about the home and family and how home and family relate to political issues. *Example*: Real family values mean that your parents, as they age, don't have to sell their home or mortgage their future to pay for health care or the medications they need.
- Avoid the usual mistakes. Remember, don't just negate the other person's claims; reframe. The facts unframed will not set you free. You cannot win just by stating the true facts and showing that they contradict your opponent's claims. Frames trump facts. His frames will stay and the facts will bounce off. Always reframe and fit the facts to *your* frame.
- If you remember nothing else about framing, remember this: *Once your frame is accepted into the discourse, everything you say is just common sense.* Why? Because that's what common sense is: reasoning within a commonplace, accepted frame.
- Never answer a question framed from your opponent's point of view. Always reframe the question to fit your values and your frames. This may make you uncomfortable, since normal discourse styles require you to directly answer questions posed. That is a trap. Practice changing frames.
- Be sincere. Use frames you really believe in, based on values you really hold.

- A useful thing to do is to use rhetorical questions: *Wouldn't it be better if . . . ?* Such a question should be chosen to presuppose your frame. *Examples*: Wouldn't it be better if we could fix the potholes in the roads and the bridges that are crumbling? Or, wouldn't we all be better off if everybody with diseases and illnesses could be treated so that diseases and illnesses wouldn't spread? Or, wouldn't it be better if all kids were ready for school when they went to kindergarten?

- Stay away from set-ups. Fox News shows and other rabidly conservative shows try to put you in an impossible situation, where a conservative host sets the frame and insists on it, where you don't control the floor, can't present your case, and are not accorded enough respect to be taken seriously. If the game is fixed, don't play. And if you do play, reframe and don't be a patsy.

- Tell a story. Find stories where your frame is built into the story. Build up a stock of effective stories.

- Always start with values, preferably values all Americans share such as security, prosperity, opportunity, freedom, and so on. Pick the values most relevant to the frame you want to shift to. Try to win the argument at the values level. Pick a frame where your position exemplifies a value everyone holds—like fairness. Example: Your uncle says, "We need right-to-work laws. Unions are corrupt and run by thugs. They force you to join and just take your money." Response: "Unions make you free—free from being a slave to a company. Without a union, you have to take whatever wage the company offers, often with no pension or medical care, with no constraints on hours or scheduling, and no guaranteed overtime pay. I wouldn't want to be a slave to a company I work for. I want to be able to eat dinner with my family and have weekend time with my kids. Unions created weekends. People used to have to work six-day weeks for less pay than they get now. Unions created eight-hour days, when people used to work ten or twelve for no more pay. Unions put you on an even basis with the company. I want to be paid fairly, treated fairly, be respected in the company where

I work, and feel good about the company. I'm not interested in being a slave. Whatever I pay to a union I more than make up for with pay from my job."

- Be prepared. You should be able to recognize the basic frames that conservatives use, and you should prepare frames to shift to. My website, www.georgelakoff.com, posts analyses of frame shifting. *Example*: A tax cut proponent says, "We should get rid of taxes. People know how to spend their money better than the government." *Reframe*: "The government has made very wise investments with taxpayer money. Our interstate highway system, for example. You couldn't build a highway with your tax refund. The government built them. Or the Internet, paid for by taxpayer investment. You could not make your own Internet. Most of our scientific advances have been made through funding from the National Science Foundation and the National Institute of Health—great government investments of taxpayer money. Computer science was developed with taxpayer money, so was the satellite system, so were the chips in our cell phones, so were the wonder drugs we need. No matter how wisely you spent your own money, you'd never get those scientific and medical breakthroughs. And how far would you get hiring your own army with your tax refund?"

- Use wedge issues, cases where your opponent will violate some belief he holds no matter what he says. Student debt is a good example. Ask if he believes in equality of opportunity and an opportunity society, which conservatives have continuously argued for (as opposed to "equality of outcome.") *Reframe*: "Many poor students with talent can only go to college if they get a government loan. But those loans cost between 6 percent and 12 percent interest and leave students with a mountain of debt that many cannot afford. The income from that debt yields profit for the government that is scheduled to be funneled into the general fund for many years into the future. Elizabeth Warren has proposed lowering the student debt interest rate to an affordable 3.86 percent, still giving the government some profit, while making up the profit lost to the government by plugging tax loopholes that allow the rich to avoid paying taxes. The students would then go to college, get out without a mountain of debt, and then be able to use the money they earn—not to pay off the government loans, but to get married, buy homes, and have kids, spending that money in the economy and boosting the economy and creating jobs. Do you want equality of opportunity with the poor able to afford college loans and boost the economy or do you want to protect unfair tax loopholes for billionaires and kill off equality of opportunity?"

- An opponent may be disingenuous if his real goal isn't what he says his goal is. Politely point out the real goal, then reframe. *Example*: Suppose he starts touting smaller government. Point out that conservatives don't really want smaller government. They don't want to eliminate the military, or the FBI, or the Treasury and Commerce Departments, or the nine-tenths of the courts that support corporate law. That is big government that they like. What they really want to do away with is social programs—programs that invest in people, that help people to help themselves. Such a position contradicts the values the country was founded on—the idea of a community where people pull together to help each other. From John Winthrop on, that is what our nation has stood for.

- Your opponent may use language that means the opposite of what she says, called Orwellian language. Realize that she is weak on this issue. Use language that accurately describes what she's talking about to frame the discussion your way. *Example*: Suppose she cites the "Healthy Forests Initiative" as a balanced approach to the environment. Point out that it should be called "No Tree Left Behind" because it permits and promotes clear-cutting, which is destructive to forests and other living things in the forest habitat. Use the name to point out that the public likes forests, doesn't want them clear-cut, and that the use of the phony name shows weakness on the issue. Most people want to preserve the grandeur of America, not destroy it. Don't you?

- Remember once more that our goal is to unite our country behind our values, the best of traditional American values. Right-wing ideologues need to divide our country via a nasty cultural civil war. They need discord and shouting and name calling and put-downs. We win with civil discourse and respectful cooperative conversation. Why? Because it is an instance of the nurturant model at the level of communication, and our job is to evoke and maintain the nurturant model.

Those are a lot of guidelines. But there are only four really important ones:

Show respect
Respond by reframing
Think and talk at the level of values
Say what you believe

★ ACKNOWLEDGMENTS ★

Each morning, my wife, Kathleen Frumkin, gets to the morning paper before I do and homes in unerringly on the deep and often hidden implications of the main political issues of the day. Much of what appears in this book is a response to her insights.

Pamela Morgan edited the first version of the talk that appears as chapter 1. She also helped me work through many of the issues discussed throughout the first edition.

Don Hazen, editor of *AlterNet*, had the idea for the first version of this book and did a great deal to make it possible. He has been a constant source of important questions and of help, intellectual and otherwise, in many ways.

Elisabeth Wehling has helped me work through many ideas, both as student and colleague.

Many of the ideas discussed here arose in discussions with those connected to the Rockridge Institute: Larry Wallack, Peter Teague, Bruce Budner, Eric Hass, Sam Ferguson, Joe Brewer, Jason Patent, Dan Kurtz, Katherine Allen, Alyssa Wulf, David Brodwin, Fred Block, Carole Joffe, Jerome Karabel, Kristen Luker, Troy Duster, Ruth Rosen, Jessica DiCamillo, Melinda Franco, Jonathan Frank, Cathy Lenz, Jodi Short, and Jessica Stites.

Other friends who have contributed ideas in discussions include George Akerlof, Don Arbitblit, Paul Baer, Peter Barnes, Joan Blades, Wes Boyd, Tony Fazio, David Fenton, Paul Hawken, Arianna Huffington, Dan Kammen, the late Anne Lipow, Ted Nordhaus, Geoff Nunberg, Karen Paget, Robert Reich, Lee Rosenberg, the late Jon Rowe, Guy Saperstein, Michael Shellenberger, Steve Silberstein, Daniel Silverman, Glenn Smith, George Soros, Alex Steffen, Deborah Tannen, Adam Werbach, Lisa Witter, Rebecca Wodder, and Richard Yanowitch.

And finally, a toast to the Father of Frame Semantics, my longtime Berkeley colleague and one of the greatest linguists ever, the late Charles J. Fillmore, who first introduced me to the political importance of his work. His name should be honored by everyone who has become aware of the importance of framing.

★ ABOUT THE AUTHOR ★

BART NAGEL

George Lakoff is the country's leading expert on the framing of political discourse and one of the world's most renowned linguists and cognitive scientists. He is the author of numerous books on politics—including *Don't Think of an Elephant!*, *The Political Mind*, *Moral Politics*, *Thinking Points*, *The Little Blue Book* (with Elisabeth Wehling), and *Whose Freedom?*—as well as numerous books on language and the mind.

Lakoff has consulted with the leaders of hundreds of advocacy groups on framing issues, lectured to large audiences across the country, run dozens of workshops for activists, spoken regularly on radio talk shows and TV shows, addressed the policy retreats and the caucuses for both the Senate and the House Democrats, consulted with progressive pollsters and advertising agencies, been interviewed at length in the public media, and continues to do extensive research on the framing of public discourse.

Currently Distinguished Professor of Cognitive Science and Linguistics at the University of California at Berkeley, Lakoff is a founder of the fields of cognitive science and cognitive linguistics and has previously taught at Harvard and the University of Michigan and was a fellow at the Center for Advanced Study in the Behavioral Sciences at Stanford. For more than two decades, he was codirector of the Neural Theory of Language Project at the International Computer Science Institute at Berkeley. He also spent more than a decade as senior fellow at the Rockridge Institute, a nonpartisan think tank, and has served on the international advisory board of Prime Minister José Zapatero of Spain, on the science board of the Santa Fe Institute, as president of the International Cognitive Linguistics Association, and on the governing board of the Cognitive Science Society, where he is now a fellow of the society. He has lectured at major universities in dozens of countries around the world. His current technical research is on the theory of how the neural circuitry of the brain gives rise to thought and language.

His blogs appear regularly on his website (www.georgelakoff.com) and at *The Huffington Post, Truthout, AlterNet, Common Dreams,* and *Daily Kos.*

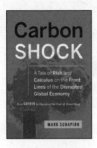

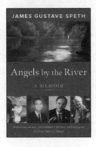

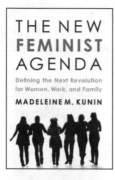

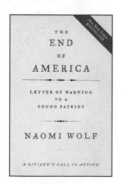